Sustainable Aviation

Series Editors

T. Hikmet Karakoc, Eskisehir Technical University, Eskisehir, Türkiye

C. Ozgur Colpan, Department of Mechanical Engineering, Dokuz Eylül University, Buca, Izmir, Türkiye

Alper Dalkiran, School of Aviation, Süleyman Demirel University, Isparta, Türkiye

The Sustainable Aviation book series focuses on sustainability in aviation, considering all aspects of the field. The books are developed in partnership with the International Sustainable Aviation Research Society (SARES). They include contributed volumes comprising select contributions to international symposiums and conferences, monographs, and professional books focused on all aspects of sustainable aviation. The series aims at publishing state-of-the-art research and development in areas including, but not limited to:

- Green and renewable energy resources and aviation technologies
- Aircraft engine, control systems, production, storage, efficiency, and planning
- Exploring the potential of integrating renewables within airports
- Sustainable infrastructure development under a changing climate
- Training and awareness facilities with aviation sector and social levels
- Teaching and professional development in renewable energy technologies and sustainability

* * *

Selim Gürgen
Editor

Advances in Manufacturing for Aerospace Alloys

Springer

Editor
Selim Gürgen
Department of Aeronautical Engineering
Eskişehir Osmangazi University
Eskişehir, Türkiye

ISSN 2730-7778 ISSN 2730-7786 (electronic)
Sustainable Aviation
ISBN 978-3-031-64454-2 ISBN 978-3-031-64455-9 (eBook)
https://doi.org/10.1007/978-3-031-64455-9

This Springer imprint is published by the registered company Springer Nature Switzerland AG
The registered company address is: Gewerbestrasse 11, 6330 Cham, Switzerland

Preface

Manufacturing in the aerospace industry is a specific area that requires special care, as aerospace alloys are highly resilient to challenging conditions during their service. Conventional manufacturing methods are mostly insufficient for shaping aerospace alloys, prompting the adaptation of nonconventional techniques for these materials. The book *Advances in Manufacturing for Aerospace Alloys* focuses on advanced manufacturing operations in the aerospace industry, such as vibration-assisted machining, electrical discharge machining, additive manufacturing, etc. In addition to conventional operations, this book explains nonconventional processes that have attracted significant attention from the aerospace industry in recent years. As a result, it compiles reference knowledge in this specific field. Furthermore, the book covers not only the theoretical aspects of the subjects but also their practical and applied sides, which are crucial in the field of engineering.

Eskişehir, Türkiye Selim Gürgen

Contents

About the Editor

Selim Gürgen is an Associate Professor with the Department of Aeronautical Engineering at Eskişehir Osmangazi University, and the Director of Eskişehir Vocational School. He is an Executive Board Member of the ESOGU Aerospace Technologies Research Center. He was a visiting researcher at the University of Wollongong for a project on smart materials. He was at the University of Aveiro on natural composites, and worked as a researcher with the School of Transportation at Anadolu University for four years. Before his academic life, he worked as a manufacturing engineer in a joint venture of KMWE-Dutch Aero. Dr. Gürgen has contributed to several industrial and research projects in collaboration with major organizations such as Fiat Automobiles, Turkish Aerospace Industries (TAI), and the Turkish Air Force. His projects were funded by the Scientific and Technological Research Council of Turkey (TÜBİTAK) and the European Union (EU). He has many publications in refereed international journals and conferences. He serves as an editorial member for several international journals.

Chapter 1
Turning Operations of Aerospace Alloys

Jalal Joudaki, Mehdi Safari, and Ricardo Alves de Sousa

1.1 Introduction

The turning process is a traditional method used for fabricating cylindrical parts. In this process, the workpiece is clamped in the chuck and rotates at a constant rotational speed. A hard tool approaches the rotating workpiece and begins material removal. There are two main turning operations: straight turning and facing. In straight turning, the tool travels along the longitudinal axis of the workpiece, while in facing operation it travels along the radial direction. The primary process parameters of turning are rotational speed, depth of cut, and feed rate. The cutting velocity is determined based on the tool's ability to withstand the applied force, transfer the generated heat during chip formation, and the lathe machine's capability to provide the required rotational energy and structural rigidity.

The turning process involves material removal through the application of a mechanical load, leading to plastic deformation. As a result, significant heat is generated by plastic deformation and friction loads in the contact zones of the tool, workpiece, and chip. Therefore, the turning process can be considered a thermomechanical phenomenon, wherein the stresses experienced can be as high as the ultimate tensile strength of the material. The elastoplastic material behavior affects various aspects of the process, including tool wear, surface roughness, induced residual stresses, and chip thickness and morphology.

Various materials, such as aluminum alloys, titanium alloys, magnesium alloys, steels, nickel-based and cobalt-based superalloys, polymers, composites, and even woods, are utilized in the aerospace industry. Mouritz [1] extensively discussed

J. Joudaki · M. Safari (✉)
Department of Mechanical Engineering, Arak University of Technology, Arak, Iran

R. A. de Sousa
Center for Mechanical Technology and Automation, Department of Mechanical Engineering, University of Aveiro, Aveiro, Portugal

S. Gürgen (ed.), *Advances in Manufacturing for Aerospace Alloys*, Sustainable Aviation, https://doi.org/10.1007/978-3-031-64455-9_1

different types of materials used in aerospace applications. According to this work, turning operations must be tailored by considering the mechanical and thermal properties of each material. The purpose of this chapter is to discuss the different aspects of turning operations in aerospace materials. It is essential to examine the diverse materials used for fabricating aerospace structures, including airplanes, helicopters, and shuttles. Rotor parts, vanes, nozzles, engine components, structural and fabrication parts, avionics components, and drone parts are among the examples of aerospace parts typically fabricated using turning processes. Due to the simultaneous requirement of high strength and low density in aerospace applications, aluminum alloys are particularly desirable materials for this industry. Common parts made of aluminum alloys, specifically AA2024, AA6061, and AA7075, include aircraft fittings, fuselage parts, shafts, and certain missile components.

The aerospace industry stands out as a leading sector among various life domains, constantly introducing new technologies, materials, and processes. Figure 1.1 illustrates the main concerns and constraints in material selection in aerospace industries. Achieving considerably higher mechanical properties is essential to support higher velocities with safe operation, while reducing weight is equally crucial for decreasing fuel consumption. However, a significant challenge lies in preparing the material's chemical composition. Some alloying elements that enhance strength may not be compatible (soluble) with the base metal or may form brittle compositions. These challenges have led to the utilization of a variety of materials in aerospace industries, adhering to specific design criteria.

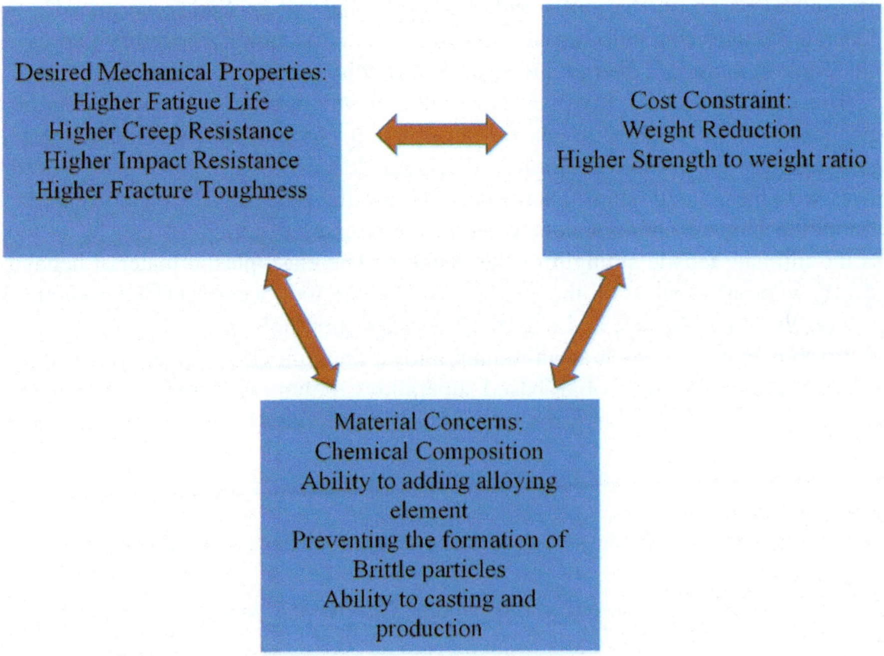

Fig. 1.1 Main domains of material selection challenge in aerospace industries

Another significant challenge in metal cutting is the selection of cutting fluids for machining operations. Dry turning has emerged as an environmentally friendly solution for machining many high-strength and very poorly machinable materials [2]. Among the conventional cutting fluids, water-based cutting fluids are widely used. However, water alone cannot provide sufficient lubrication and cooling due to its corrosive nature and limited lubricating properties. To overcome this limitation, additional ingredients (usually 15–20) are added to improve lubrication. Water-based cutting fluids typically cost between US$5 and US$20 per gallon and are diluted at ratios ranging from 1:5 to 1:20, depending on the type of the fluid.

One of the challenges associated with water-based cutting fluids is their thermodynamic instability, requiring them to be changed after circulating for a specified period. This creates an environmental problem as the used cutting fluid cannot be easily and economically recycled due to its inorganic ingredients. An alternative solution is the use of oil-based cutting fluids and transitioning toward minimum quantity lubrication (MQL). Mineral oils, while offering good lubricity, have limited cooling capacity. MQL involves adding a small amount of oil (typically 5–60 mL/h) to compressed air and spraying it onto the workpiece to provide effective lubrication. Other strategies for cutting fluid optimization include minimum quantity cooled lubrication (MQCL), liquid nitrogen cryogenic cooling, supercritical carbon dioxide ($SCCO_2$) cooling, and water-based cutting fluids formulated with bio-oils [3].

1.2 Process Parameters in Turning

Turning is both a simple and a complex process in manufacturing. Among its variables, cutting speed stands as the foremost parameter, influencing and determining the selection of other variables. Figure 1.2 illustrates the interaction between rotational speed and other process parameters, with feed rate and cutting depth being significant factors. Tool life emerges as a critical factor in selecting and balancing these parameters. Increasing the cutting speed results in reduced machining time and higher production rates. However, elevated cutting speeds lead to heightened temperatures, accelerating wear on the tool's flank surface. To mitigate this, manufacturers often employ high-pressure wet lubrication for cooling the cutting zone. Failure to adequately cool the cutting zone can significantly reduce tool life, necessitating frequent tool changes. This, in turn, increases tool costs, thereby diminishing profits despite the increased production rate. Hence, maintaining an optimal tool life while managing tool-changing costs is pivotal for achieving higher production numbers profitably. Eq. 1.1, proposed by Marksberry et al. [4], offers a means to determine tool life under various cutting conditions. This equation was developed based on Taylor's fundamental eq. ($VT^n = C$).

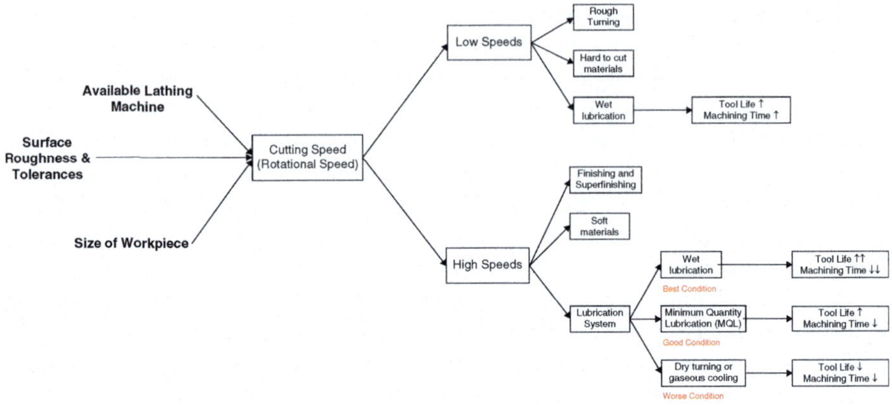

Fig. 1.2 The criteria for the selection of cutting speed in turning process

$$\frac{T}{T_R} = \frac{k}{f^{n_1} d^{n_2}} \left(\frac{V^R}{V}\right)^{\left(\frac{1}{n_c}\right)\left(\frac{1}{N_{\mathrm{NDM}}}\right)} \tag{1.1}$$

where

T is the tool life,
T_R is the reference tool life for 1 minute of turning,
V denotes the cutting speed,
V_R is the reference cutting speed,
K, n_1, and n_2 are empirical constants used to determine the effects of cutting speed, feed rate, and depth of cut, respectively,
n_c represents the effect factor of tool coating, and,
N_{NDM} represents the contribution of the type of coolant/lubrication system in the turning process.
Moving from wet turning to near-dry machining (NDM) reduces the N_{NDM} factor.

Traditional cutting fluids, typically derived from neat cutting oils or water-soluble fluids, serve the purpose of removing chips from the tool's rake face and dissipating the heat generated during cutting. Water has been a common base for cutting fluids due to its excellent thermal capacity. As depicted in Fig. 1.2, low cutting speeds are employed for roughing operations, resulting in significant heat generation during turning. In such scenarios, the use of flooded lubrication enhances tool life, albeit at the expense of longer machining times, leading to lower production rates. Conversely, high cutting speeds (often combined with low feed rates and depths of cut) can yield surfaces with excellent roughness and tight tolerances. This approach reduces machining time while comparatively minimizing the volume of the chips produced.

1.3 Turning of Aluminum Alloys

Aluminum is a highly desirable material for turning and machining operations due to its ability to achieve good surface roughness and low induced residual stresses across various grades of aluminum alloys. Among these alloys, AA2024, AA6061, AA6063, and AA7075 are particularly notable for their machinability, especially when in heat-treated conditions. Chip breakers are often employed to improve machinability by facilitating the formation of continuous chip morphology during turning.

Experimental tests have revealed that increasing cutting speed results in decreased surface hardness and induced residual stresses. This phenomenon is attributed to the enhanced rate of heat generation and dissipation over the turned surface. Conversely, increasing the depth of cut leads to higher residual stresses and machining forces, including cutting force, feed force, and thrust force [5].

The rate of decrease in machining forces is observed to be relatively slow with increasing cutting speed [6]. Table 1.1 provides a list of references related to the turning of aluminum alloys, highlighting the wide range of process parameters utilized in various studies. Further investigations are warranted for each specific case of part manufacturing to optimize process parameters and enhance machining efficiency.

In certain studies, refined vegetable oil has been utilized as a cooling/lubricating agent in the turning process, particularly notable in the cutting of AA6063-T5 [8]. Findings suggest that refined vegetable oil exhibits a reduction in frictional force and normal force compared with mineral oil. In addition, a decrease in the contact length between the chip and the tool's rake face is observed, attributed to the reduction in the frictional and normal forces. Despite the favorable characteristics of refined vegetable oil, challenges arise due to its transition temperature. While part of the heat generated results from friction between the tool's rake face and the deformed chips, the majority stems from high plastic deformation in the shear plane or shear zone. Effective cooling/lubricating fluids should minimize friction force and heat generation in the shear plane. Mineral oil excels at heat dissipation but lacks lubricity, with its properties minimally affected by increases in temperature. Conversely, refined vegetable oil demonstrates good lubricity up to temperatures of around 200 °C but possesses limited thermal capacity. Consequently, in the cutting zone, temperatures rise, leading to a decline in the lubricity properties of refined vegetable oil beyond its transition temperature. As a result, at higher temperatures, such as those associated with higher cutting speeds (e.g., 400 m/min), refined vegetable oil and mineral oil exhibit a similar behavior.

Aluminum, being a soft material, presents a challenge in turning due to the tendency of the deformed chip to stick onto the rake face of the tool. The coolant fluid plays a crucial role in mitigating this issue, as it influences the quantity of adhering material on the tool. Increasing the cutting speed results in more heat generation, leading to greater thermal softening of the chips and subsequently increasing adhesion between the chip and the tool surfaces (flank face, rake face,

Table 1.1 Cutting conditions in turning of aluminum alloys

References	Material	Cutting speed (m/min)	Feed rate (mm/rev)	Depth of cut (mm)	Tool	Other details
Sreejith [7]	AA6061	50–400	0.15	1.0	ISO CNGA 120408 T01020 WG	Flood coolant, dry cutting, and MQL
Itoigawa et al. [8]	AA6063-T5	100, 200, 400	0.1	1.5	Carbide (G10)	0.11 μm
Niknam and Jalali [9]	AA6061-T6 AA7076-T6	79, 116, 163, 206, 442, 660	0.0508, 0.1016, 0.1524, 0.2032, 0.254	1	KC5410-coated carbide insert	MQL Full-factorial design of experiment (DOE)
Sa'nchez et al. [10]	AA2024 and AA7050	40–200	0.05–0.3	2	ISO KCMW 11 T308 FN M	Dry BUE observed
Surya Sundara Rao and Allamraju [5]	AA7075	50, 100, 200, 300	0.05	0.3, 0.6, 0.9	carbide insert coated with TiB_2	0.57–0.72 μm roughness
Haddag et al. [6]	AA2024-T351	60–500	0.1, 0.3	4	Uncoated WC-Co carbide insert type K4	Arbitrary Lagrangian–Eulerian finite element model is developed
Ni et al. [11]	AA1100	0.6	0.3	–	Kennametal CNGA-432	–
Eapen et al. [12]	AA6063	70, 110, 175	0.2, 0.315, 0.4	2.5	General purpose insert	Parts hold at −77 °C and dry turned
Rotella [13]	AA7075-T6	90, 120, 150	0.15	0.05	Uncoated insert	Dry Cryogenic MQL High-pressure air jet 0.2–0.6 μm roughness
Yücel et al. [14]	AA2024 T3	300, 400, 500	0.1, 0.2, 0.3	1	ISO VCGT 160404 FL	Dry MQL Nanofluid MQL 0.8–7.6 μm
Junge et al. [15]	AA2017 T4	50, 300, 550	0.04	0.4	CVD-coated insert	Dry Air cooling Flood MQL 0.13–0.32 μm

and clearance surface) [8]. However, experiments on flank wear reveal varied trends. Dry turning tends to induce higher wear on the flank face, whereas employing minimal quantity lubrication (MQL) or flooded lubrication yields similar, lower levels of flank wear. Nonetheless, it remains evident that flank wear escalates with increasing cutting speed. Niknam et al. [9] investigated the impact of process parameters on average surface roughness (R_a). Their findings indicate that the most influential factors are the feed rate and the square of the feed rate. While the flow rate of lubricant and cutting speed also play roles, their effects are relatively limited. A notable linear correlation was observed between surface roughness and chip thickness.

In turning operation of aluminum alloys, the formation of built-up edge (BUE) is a significant topic of discussion. Tool life, as mentioned earlier, can be evaluated based on various criteria, one of which is tool wear. Tool wear is closely related to the adhesion of chemical elements from the tool, which are intended to reduce friction in the coating layer, to the chips. In addition, it involves the adhesion of chips to the rake face of the tool. In dry turning conditions, the friction coefficient typically ranges between 0.55 and 0.70. The sticking of chips to the rake face is initially unstable. However, as the BUE grows to a certain size, the cutting force causes its removal, resulting in adhesion wear on the rake face. Experimental studies have identified two types of adhesion of machined materials to the rake face:

(i) Built-up edge (BUE): sticking of removed materials to tool edge.
(ii) Built-up layer (BUL): sticking of removed materials to rake face.

Various studies have investigated the mechanisms behind the formation of built-up edge and built-up layer [10–16]. The primary factor influencing their formation is the rise in temperature. In dry turning processes involving aluminum alloys and similar soft materials, the temperature at the contact surface of the tool (rake face) and chips can reach up to 750 °C. This temperature elevation leads to the melting of aluminum material, resulting in the formation of an extremely thin layer on the rake face due to the flow of melted material. Subsequently, the compressive cutting force causes the welded attachment of this cooled layer to the tool surface. As the temperature rises, the thickness of the built-up layer gradually increases. Meanwhile, material from the built-up edge, located near the tool edge, is mechanically adhered and removed. Therefore, the formation of built-up edge is essentially the result of the mechanical adhesion processes that lead to the formation of built-up layer. The thickness of built-up edge increases until it reaches a critical thickness, at which point it extends over the built-up layer. Figure 1.3 shows the formation of built-up edge and built-up layer during the turning process of soft materials. One effective method for mitigating the formation of built-up edge is the use of nanofluids. For instance, Yücel et al. [14] utilized molybdenum disulfide (MoS_2) nanoparticles to reduce the maximum temperature by approximately 22%, prevent built-up edge formation, and enhance surface roughness.

Rotella [13] conducted a study to assess the impact of cutting speed and cooling systems on the high cycle fatigue behavior of aluminum alloys. The findings revealed that, at higher cutting speeds, there was a decrease in surface roughness

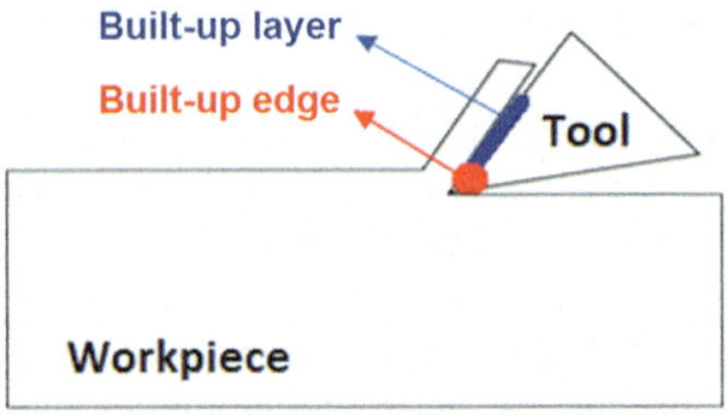

Fig. 1.3 Formation of built-up edge and built-up layer in turning

and average grain size, which subsequently inhibited fatigue crack initiation. In metal cutting processes, the application of severe stress beneath the surface induces grain refinement, resulting in increased surface hardness. This phenomenon is particularly significant for aluminum alloys such as AA7075, where a fatigue life cycle exceeding 10^6 cycles can be achieved, even at a stress amplitude of 175 MPa, a critical requirement in aerospace applications.

Notably, the study demonstrated that the combination of high-speed machining (150 m/min) and the utilization of a cryogenic cooling system yielded the best fatigue life cycle results. The cryogenic cooling system, employing two 2-mm-diameter nozzles of liquid nitrogen at 12 bar pressure, facilitated higher cooling rates, effectively preventing the recovery phenomena following dynamic recrystallization. Consequently, the grain size remained finer, leading to retardation in crack initiation. Conversely, dry turning resulted in larger grain sizes and higher surface roughness, contributing to a decrease in fatigue life. These findings underscore the importance of selecting appropriate cutting parameters and cooling systems to optimize the fatigue behavior of aluminum alloys in aerospace applications.

1.4 Turning of Titanium Alloys

Turning titanium alloys presents unique challenges compared with aluminum alloys due to the formation of segmented chip morphologies, as opposed to continuous chips typically observed in aluminum alloy turning. This segmentation of chips introduces variations in the applied load to both the tool and the workpiece, resulting in vibrational movements during material removal. These vibrations, coupled with the cutting force, can significantly impact surface roughness and induce residual stresses in the workpiece.

For further insights into the turning of titanium alloys, Table 1.2 provides a compilation of references relevant to this machining process. These references offer valuable information and insights into addressing the specific challenges associated with turning titanium alloys, aiding in the development of effective machining strategies and optimization techniques for improved machining outcomes.

Table 1.2 Cutting conditions in turning of titanium alloys

References	Material	Cutting speed (m/min)	Feed rate (mm/rev)	Depth of cut (mm)	Tool	Other details
Settineiand Faga [17]	TA48	50, 90, 130	0.25	1	PVD-coated ISO K05 WC-Co inserts	MQL ~0.32 μm R_a
Wang et al. [18]	Ti6Al4V	60, 75, 90, 105, 120	0.10, 0.15	3	ISO TNGA-160404 Uncoated WC	Flood coolant, dry cutting, and MQL
Deiab et al. [19]	Ti6Al4V	90, 120	0.1, 0.2	0.8	CCMT 09 T3 08-KM H13A	Dry MQL MQCL Cooled air Flood Cryogenic 0.8–2 μm R_a
Palanisamy et al. [20]	Ti6Al4V	75	0.25	2	CNMG120408	Coolant at high pressure (90 bar)
Abboud et al. [21]	Ti6Al4V Ti6Al2Sn4Zr6Mo	20–90	0.05–0.25	0.1–0.3	Uncoated WC tool	Flood lubrication
Rubio et al. [22]	Ti6Al4V	1.3, 2.4	0.11, 0.2	–	TP100 TP200 TP1000	Dry MQL Compressed Air 0.8–1.6 μm R_a
Airao et al. [23]	Ti6Al4V	50	0.2	0.3	WC with TiCN-Al2O3-TiN coating	Dry and Vegetable oil cutting fluid Ultrasonic assisted turning 0.8–1.7 μm R_a
Swain et al. [24]	Ti6Al4V	75	0.1	0.4	Uncoated carbide CNMG 1020408	Dry MQL ~0.8 μm R_a
Da Silva et al. [25]	Ti6Al4V	175, 200, 230, 250	0.15	0.5	Polycrystalline diamond	Flood cooling and high-pressure cooling

Rubio et al. [22] conducted a study on the surface roughness variation during the turning of 50-mm Ti6Al4V bars. Using a design of experiments (full factorial approach), they investigated the effects of process parameters, particularly cutting speed and feed rate. The cutting speed was varied at two levels (500 and 900 rpm), while feed rate and cooling system were also considered. Their results revealed that surface roughness increased with higher feed rates and spindle speeds, regardless of the type of cooling system employed. The maximum surface roughness was observed at a feed rate of 0.2 mm/rev and a spindle speed of 925 rpm, while the minimum surface roughness was achieved at a feed rate of 0.11 mm/rev and a spindle speed of 500 rpm. Interestingly, at low feed rates, the difference in surface roughness between 500 and 925 rpm spindle speeds was negligible. Furthermore, the type of cooling system had a significant effect on surface roughness. For instance, at a spindle speed of 500 rpm, transitioning from dry cutting to MQL and then compressed air resulted in a decrease in surface roughness. However, at a spindle speed of 925 rpm, there was a notable increase in surface roughness with MQL and compressed air cooling methods. Based on their findings, the authors recommended lower spindle speeds and feed rates for repairing and maintaining aeronautical and aerospace parts. In addition, they suggested that using a cladded tool with TiCN and Al_2O_3 coating could enhance surface roughness, particularly at high spindle speeds. Moreover, they observed that chip thickness depended on the generated heat during turning, with thinner chips produced at lower cutting speeds and feed rates. Strategies such as high-pressure air jet and flooded lubrication were effective in reducing heat generation, while MQL and dry cutting were less effective, particularly at high cutting speeds and feed rates [18].

Airao et al. [23] explored the integration of conventional turning with ultrasonic vibration, facilitated by an ultrasonic horn applying waves at a frequency of 20 kHz and an amplitude of 10 μm. Their study aimed to investigate the impact of ultrasonic vibration on surface roughness during dry turning and turning with lubrication using a vegetable oil emulsion (10% emulsion of canola oil in mineral water). Their results demonstrated that the application of ultrasonic vibration led to a substantial reduction in surface roughness, achieving a 42% decrease under dry turning conditions. In addition, lubrication with the vegetable oil emulsion reduced surface roughness by approximately 28% compared with dry turning conditions alone. The combination of vegetable oil lubrication and ultrasonic-assisted turning yielded the best surface roughness, with values around 0.8 μm R_a. Furthermore, the application of ultrasonic vibration resulted in an increase in surface hardness beneath the machined specimen surface, with a thickness of approximately 70 μm.

A similar investigation was conducted by Lotfi et al. [26] on American Iron and Steel Institute (AISI) 304 stainless steel. Their study also found that ultrasonic vibration reduced both the thrust force and cutting force, along with decreasing the tool–chip contact length. This reduction in contact length was attributed to an increase in the shear angle induced by ultrasonic vibration. Consequently, the generation of heat in the contact zone decreased, leading to lower temperatures and ultimately enhancing tool life.

In titanium parts subjected to turning processes, the combination of high yield strength and ultimate tensile strength, along with the considerable forces applied during material removal, leads to elevated temperatures exceeding 700 °C. This high temperature environment can induce phase transformations in the material. The formation of residual stresses in titanium-turned parts is influenced by three primary factors: mechanical loads, thermal loads, and phase transformations. These factors directly contribute to the residual stress distribution within the material, which in turn can impact the performance of the parts. Residual stresses induced during turning processes in titanium parts are typically compressive in nature. These compressive residual stresses have a beneficial effect on the fatigue life of the machined parts, enhancing their durability. Several studies by researchers such as Bin Shi et al. [27], Bai et al. [28], Abboud et al. [21], and Li et al. [29] have investigated the distribution of residual stresses in titanium alloys. It is important to note that in titanium alloys, unlike aluminum alloys, the grain size plays a dominant role in determining fatigue life. In addition, surface roughness is a significant factor influencing crack initiation and propagation in titanium alloys. Therefore, achieving appropriate residual stress distribution and surface finish is crucial for ensuring the optimal performance and longevity of titanium parts subjected to turning processes [13].

1.5 Conclusions

In conclusion, the turning operation stands as a cornerstone in the manufacturing processes in the aerospace industry, particularly for crafting symmetrical components vital for various rotating machines and engines. The careful selection of process parameters remains paramount, given its multifaceted impact on factors ranging from tool cost to surface quality, which significantly influences crucial mechanical properties such as fatigue life. This study underscores the significance of understanding the interplay between cutting forces, heat generation, and cooling rates in shaping surface roughness and microstructure, particularly in materials renowned for their high strength-to-weight ratios, such as aluminum and titanium alloys. As researches continue to reveal the intricacies of turning operations, particularly concerning parameter optimization and the mitigation of challenges like built-up edge and built-up layer formation, the integration of advanced cooling systems such as MQL and high-pressure cooling systems emerges as a pivotal strategy for enhancing manufacturing efficiency and product quality in aerospace applications. Thus, this overview serves as a valuable guide for practitioners navigating the complexities of turning operations in the aerospace industry, facilitating informed decision-making and continual improvement in machining processes.

References

1. Mouritz AP. Introduction to aerospace materials. Woodhead Publishing; 2012.
2. Safari M, Joudaki J, Emadi M. Surface quality in dry machining of 55Cr3 steel bars. Int J Iron Steel Soc Iran. 2018;15(1):33–9.
3. Skerlos SJ. Cutting fluids and their environmental impact. Encycl Tribol. 2013:655–60.
4. Marksberry PW, Jawahir IS. A comprehensive tool-wear/tool-life performance model in the evaluation of NDM (near dry machining) for sustainable manufacturing. Int J Mach Tools Manuf. 2008;48(7–8):878–86.
5. Surya Sundara Rao K, Viswanath AK. Effect on micro-hardness and residual stress in CNC turning of aluminium 7075 alloy. Mater Today Proc. 2017;4(2):975–81. https://doi.org/10.1016/j.matpr.2017.01.109.
6. Haddag B, Atlati S, Nouari M, Moufki A. Dry machining aeronautical aluminum alloy AA2024-T351: analysis of cutting forces, chip segmentation and built-up edge formation. Metals (Basel). 2016;6(9):197.
7. Sreejith PS. Machining of 6061 aluminium alloy with MQL, dry and flooded lubricant conditions. Mater Lett. 2008;62(2):276–8.
8. Itoigawa F, Takeuchi D, Nakamura T, Childs THC. Experimental study on lubrication mechanism in MQL intermittent cutting process. Mach Sci Technol. 2007;11(3):355–65.
9. Niknam SA, Jalali A. Effects of lubricants and flow rates on the surface roughness and chip thickness when MQL turning of aero-engine aluminum alloys 6061-T6 and 7076-T6. Int J Adv Manuf Technol. 2020;110(7–8):2015–22.
10. Sebasti MA, Marcos M. Microstructural characterisation of material adhered over cutting tool in the dry machining of aerospace aluminium alloys. J Mater Process Technol. 2005;165:911–8.
11. Ni H, Elmadagli M, Alpas AT. Mechanical properties and microstructures of 1100 aluminum subjected to dry machining. Mater Sci Eng A. 2004;385(1–2):267–78.
12. Eapen J, Murugappan S, Arul S. A study on chip morphology of aluminum alloy 6063 during turning under pre cooled cryogenic and dry environments. Mater Today Proc. 2017;4(8): 7686–93. https://doi.org/10.1016/j.matpr.2017.07.103.
13. Rotella G. Effect of surface integrity induced by machining on high cycle fatigue life of 7075-T6 aluminum alloy. J Manuf Process. 2019;41(February):83–91. https://doi.org/10.1016/j.jmapro.2019.03.031.
14. Yücel A, Yıldırım ÇV, Sarıkaya M, Şirin Ş, Kıvak T, Gupta MK, et al. Influence of MoS2 based nanofluid-MQL on tribological and machining characteristics in turning of AA 2024 T3 aluminum alloy. J Mater Res Technol. 2021;15:1688–704.
15. Junge T, Mehner T, Nestler A, Schubert A, Lampke T. Surface properties in turning of aluminum alloys applying different cooling strategies. Procedia CIRP. 2022;108(C):246–51. https://doi.org/10.1016/j.procir.2022.03.043.
16. Gonza A, Marcos M, Carrilero MS, Bienvenido R, Sa JM. A SEM and EDS insight into the BUL and BUE differences in the turning processes of AA2024 Al—Cu alloy. Int J Mach Tools Manuf. 2002;42:215–20.
17. Settineri L, Faga MG. Nanostructured cutting tools coatings for machining titanium. Mach Sci Technol. 2008;12(2):158–69.
18. Wang ZG, Rahman M, Wong YS, Neo KS, Sun J, Tan CH, et al. Study on orthogonal turning of titanium alloys with different coolant supply strategies. Int J Adv Manuf Technol. 2009;42 (7–8):621–32.
19. Deiab I, Raza SW, Pervaiz S. Analysis of lubrication strategies for sustainable machining during turning of titanium ti-6al-4v alloy. Procedia CIRP. 2014;17:766–71. https://doi.org/10.1016/j.procir.2014.01.112.
20. Palanisamy S, McDonald SD, Dargusch MS. Effects of coolant pressure on chip formation while turning Ti6Al4V alloy. Int J Mach Tools Manuf. 2009;49(9):739–43.

21. Abboud E, Attia H, Shi B, Damir A, Thomson V, Mebrahtu Y. Residual stresses and surface integrity of Ti-alloys during finish turning—guidelines for compressive residual stresses. Procedia CIRP. 2016;45:55–8. https://doi.org/10.1016/j.procir.2016.02.069.
22. Rubio EM, Bericua A, De AB, Marín MM. Analysis of the surface roughness of titanium pieces obtained by turning using different cooling systems. Procedia CIRP. 2018;2019(79):79–84. https://doi.org/10.1016/j.procir.2019.02.015.
23. Airao J, Nirala CK, Outeiro J, Khanna N. Surface integrity in ultrasonic-assisted turning of Ti6Al4V using sustainable cutting fluid. Procedia CIRP. 2022;108(C):55–60. https://doi.org/10.1016/j.procir.2022.04.068.
24. Swain S, Kumar R, Panigrahi I, Sahoo AK, Panda A. Machinability performance investigation in CNC turning of Ti–6Al–4V alloy: dry versus iron-aluminium oil coupled MQL machining comparison. Int J Light Mater Manuf. 2022;5(4):496–509. https://doi.org/10.1016/j.ijlmm.2022.06.002.
25. Da Silva RB, MacHado ÁR, Ezugwu EO, Bonney J, Sales WF. Tool life and wear mechanisms in high speed machining of Ti-6Al-4V alloy with PCD tools under various coolant pressures. J Mater Process Technol. 2013;213(8):1459–64. https://doi.org/10.1016/j.jmatprotec.2013.03.008.
26. Lotfi M, Amini S, Ashrafi H. Theoretical and numerical modeling of tool—chip friction in ultrasonic-assisted turning. Proc Inst Mech Eng Part E: J Proc Mech Eng. 2018;233(4):1–15.
27. Shi B, Abboud E, Attia MH, Thomson V. Effect of chip segmentation on machining-induced residual stresses during turning of Ti6Al4V. Procedia CIRP. 2022;108(C):424–9. https://doi.org/10.1016/j.procir.2022.03.066.
28. Bai J, Bai Q, Tong Z, Chen G. The influence of cutting parameters on the defect structure of subsurface in orthogonal cutting of titanium alloy. J Mater Res. 2018;33(6):720–32.
29. Li G, Lu W, Huang S, Zhang X, Ding S. Analysis and prediction of residual stresses based on cutting temperature and cutting force in rough turning of Ti–6Al–4V. Heliyon. 2022;8(11):e11661. https://doi.org/10.1016/j.heliyon.2022.e11661.

Chapter 2
High-Speed Machining for Aerospace Materials

Nikita Shubin and Muhammad P. Jahan

2.1 Introduction

In the growing industry of aerospace manufacturing, aluminum-based alloys, titanium-based alloys, and nickel-based alloys are among the most widely used materials in aircraft production [1] due to their toughness, high density, good plasticity, heat, and oxidation resistance [2]. With the increasing demand for higher manufacturing speeds, increased accuracy, and better finishing qualities of more and more complex parts, high-speed machining (HSM) became one of the most used machining methods.

Based on a simple definition, HSM can be shortly described as a machining process utilizing working speeds 5–10 times higher than the ones used by conventional machining methods. However, the cutting speed ranges for HSM vary based on multiple factors such as the choice of material due to its mechanical and thermal properties, size, type and material of tools, feed rate, and machining conditions. For example, in one of the definitions, HSM for aluminum alloys is considered to start at a cutting speed of around 1500–2000 m/min and above, while HSM for nickel-based alloys begins around 200 m/min [3]. In other works, HSM is defined as the process where the cutting forces decrease with the increase in cutting speed in contrast to conventional machining where cutting forces increase with an increase in cutting speed as can be seen in Fig. 2.1 [4]. HSM can also be defined using the DN value which is the diameter of a spindle (in mm) multiplied by the spindle speed (in rpm). By such definition, the DN value of HSM is typically around 1.5–2 million. Lastly, HSM is also sometimes defined by the dominant natural frequency of the machining

N. Shubin · M. P. Jahan (✉)
Department of Mechanical and Manufacturing Engineering, Miami University, Oxford, OH, USA
e-mail: jahanmp@miamioh.edu

S. Gürgen (ed.), *Advances in Manufacturing for Aerospace Alloys*, Sustainable Aviation, https://doi.org/10.1007/978-3-031-64455-9_2

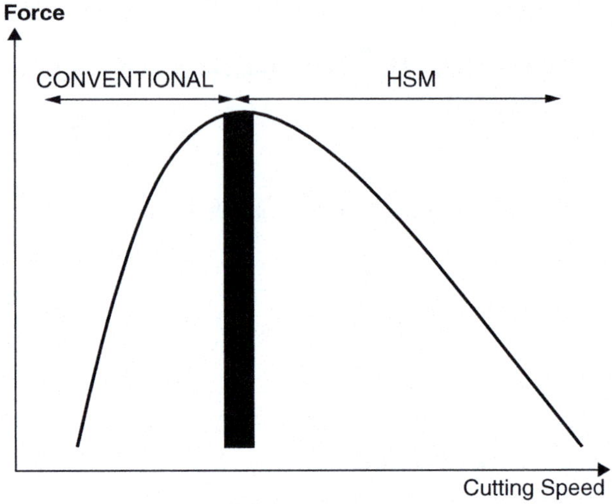

Fig. 2.1 Cutting force versus cutting speed diagram for conventional machining and HSM [4]

tool, when the angular velocity or frequency of the tool rotation reaches and goes beyond a substantial fraction (such as ¼) of the dominant natural frequency [5].

The definition of high-speed grinding differs from the other common ones because the grinding process is typically described in terms of grinding wheel surface speed in m/s, which is not a parameter typically used in definitions of HSM for milling and turning. Therefore, Li et al. [6] defined the beginning high-speed area of grinding at 80 m/s and higher for brittle materials such as Ti-6Al-4V. In addition, ultra-high-speed grinding was also defined as above 120 m/s.

Some studies use angular velocity in rpm as a measure of the cutting speed. This is especially prevalent in the studies focused on drilling. Therefore, it is necessary to convert such values to the units of m/min, which are used in the definition of HSM. This can be done with the formula given in Eq. 2.1.

$$V_c = \pi \times D_t \times R_s \tag{2.1}$$

where

V_c is the cutting speed (m/min)
D_t is the tool diameter (m), and
R_s is the rotational speed (rpm)

According to Jain et al. [7], there are multiple advantages present in HSM: significant reduction in machining times and increase in material removal rates (MRRs), higher dimensional accuracy and precision, better surface finish and quality, lower cutting forces leading to components with minimal amounts of stress, better tool life, smaller temperatures caused by better heat dissipation via chip evacuation and shorter interaction time between the tool and workpiece, and ability to work with any material on scales up to 50 nm. Therefore, HSM is often a preferred method when any of these advantages are necessary for manufacturing.

HSM is used in most of the critical applications and components in aircraft production due to its capability of achieving tighter tolerances and higher productivity compared with conventional methods. Such applications can be subdivided into airframes and aeroengines. For airframes, aluminum alloys have been generally used over the years due to their high strength-to-weight ratio, corrosion resistance, and good mechanical and thermal properties as well as ease of machinability and full recyclability. For example, more than 70% and 82% of Boeing 777 and Boing 747, respectively, are manufactured out of aluminum alloys [8]. An important role is played by Al-Zn alloys, such as AA7050 and AA7075, primarily known for their high strength and hardness. In addition, in recent years, the industry has been shifting more toward composite materials, such as metal matrix composites and carbon fiber reinforced polymers (CFRPs) for airframes. The reason for such a shift is their better mechanical properties, such as higher strength and stiffness, better fatigue and corrosion resistance, and lower weight [1]. For aeroengines, it is important to achieve higher efficiency in thermal, aerodynamic, and propulsion qualities. Therefore, materials are needed to not only be reliable and low weight, but also be able to withstand higher temperature environments and allow for better airflow. Alloys often used for such tasks include Inconel 718, Ti-6Al-4V, and ceramic matrix composites. For these alloys, it is vital to achieve precise and consistent high-quality properties, which is where HSM would be highly applicable [1].

Several important factors are always considered and attempted to be improved upon in machining of such alloys: heat generation, tool wear, surface integrity [9], as well as more recent ones such as sustainability [10]. This review briefly covers how these factors and some others are resolved with the recent advancements in HSM technology.

2.2 Machining of Aluminum Alloys

Most of the recently published papers on HSM of aluminum alloys used in aerospace covered AA7050 and AA7075. The main alloying element for these alloys is zinc, which provides them with a high strength, good fatigue and corrosion resistance, and average machinability while keeping the weight relatively small.

Aluminum alloys, as mentioned earlier, are typically utilized in the production of airframes in the aerospace industry, specifically in the manufacturing of thin-walled parts which most core airplane structures are made of. In the production of said parts, up to 95% of weight is removed from the original stock. In such applications, HSM is a common choice [11]. HSM of aluminum alloys is mostly used in milling applications while turning is applied in machining of noncritical parts, and drilling is not used often. The reasons for such decisions are that the parts produced by turning are typically considered critical components and, therefore, require better mechanical and thermal properties from materials such as Ti-6Al-4V. Furthermore, in the aerospace industry, drilling of aluminum alloy parts is mostly done with conventional techniques because of small holes with diameters around 4–10 mm typically utilized needing higher angular velocities to get into the range of HSD. In milling, however, HSM has proven to have multiple advantages over conventional methods due to better surface quality and integrity, increased productivity, faster machining time, capability of creating complex parts, and minimizing the need for additional processing [8].

2.2.1 Milling of Aluminum Alloys

Choosing proper input parameters is one of the easier and faster methods of how one can improve the productivity, tool life, and quality of the resulting part. One can vary feed rate, the depth of cut, angular velocity or the number of revolutions, tool size, and consequently cutting speed. However, these parameters have various effects that can aid or worsen the machining conditions and the quality of the resulting parts. Therefore, careful consideration needs to be given to the choice of parameters and their effects need to be known.

High cutting speed is one of the main characteristics of HSM. As expressed earlier, increasing cutting speed leads to an increase in the temperature at the tool–workpiece interface and reduction in the cutting forces of the process. Initially, during the conventional machining phase, the horizontal force (feed force) increases with the rise in the cutting speed due to the friction between the chip and the tool, while the vertical force (thrust force) decreases due to the rising temperature between the workpiece and the tool. Therefore, as the effect of the horizontal force is larger compared with the effect of the vertical force, the overall cutting force increases. As the speed continues to grow and the process enters the HSM area, the further increase in temperature causes thermal softening to occur, which changes the effective stress–strain curve, therefore increasing the shear angle and decreasing the amount of plastic deformation needed to deform the chips. This effect causes a reduction in horizontal force and marginal growth in the vertical force, which leads to an overall decrease in the total cutting force [12–15].

Cutting speed also affects the surface roughness of the material. It was shown for AA7050 that within the HSM range the roughness first tends to increase with the cutting speed up until around 4250 m/min and then begins to decrease as the cutting speed increases further as seen in Fig. 2.2 [16, 17].

The depth of cut also influences the cutting forces and temperature. As the depth of cut increases, the cutting area also becomes higher. This leads to an increase in heat generation during the process and, consequently, to rise of the temperature and the cutting forces [15]. Increasing feed per tooth leads to the removal of a larger volume of material per pass. Because of that, normal stress also increases, leading to higher cutting forces [14].

As there are many adjustable machining parameters and one can change to affect the machining process and the resulting quality of the product in different ways, choosing the proper ones becomes a balancing act. Some studies have been conducted about increasing certain output characteristics such as material removal

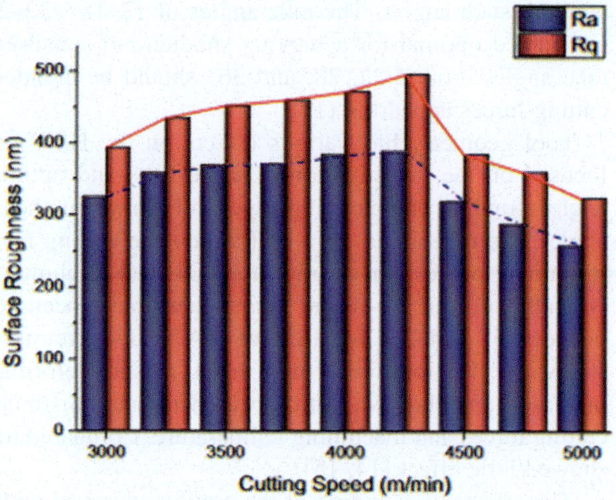

Fig. 2.2 Surface roughness changes with the cutting speed in HSM of AA7050 [16]

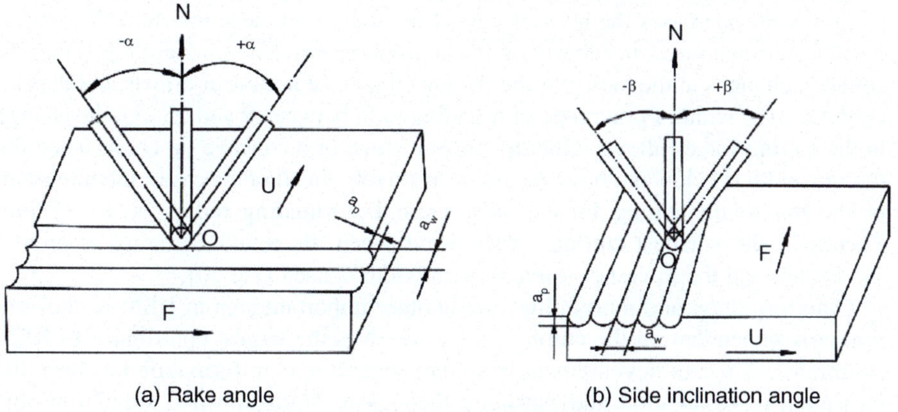

Fig. 2.3 Distinction between rake angle and side inclination angle [19]

rate (MRR), surface roughness, and surface quality. For example, Lukic et al. [18] have conducted an investigation into HSM of AA7075 thin-walled parts based on standard Taguchi orthogonal sequence with 27 experiments and multicriteria optimization with 14 multicriteria decision-making (MCDM) methods. Weighing MRR the highest, they optimized the machining parameters as follows: angular velocity of 12,000 rpm, feed of 2400 mm/min, and depth of cut of 2 mm for the wall thickness of 1 mm. This resulted in an MRR of 14,980 mm^3/min while keeping the average surface roughness at 1 μm and achieving precise dimensional surface quality (thickness, flatness, parallelism, and perpendicularity).

For ball-end mill cutting tools, there were recent developments regarding the prediction and optimization of tool axis inclination angle, specifically the rake angle while keeping the side inclination angle constant. Figure 2.3 shows the distinction between such angles. The rake angles of 12–18°, 22–26°, 30–34°, and 38–60° are considered optimal for achieving smaller and consistent cutting force. In contrast, rake angles around 22, 28, and 36° should be avoided due to unsteady and high cutting forces appearing [19].

Tool geometry has various effects on the HSM process. Recent research has focused on the investigation of the effects and optimization of rake angle, relief angle, rounding of the cutting edge, and changes in the margin width of the tool. The choice of a rake angle for a tool affects the cutting forces and temperatures during machining process. In a softer material such as aluminum, positive rake angles are typically utilized. In recent years, there have been studies of rake angle effects between 5° and 21° in HSM of AA7050. Increasing the rake angle causes the following: friction force and metal plastic deformation decrease, shear angle increases, and chip evacuation becomes easier. Such changes induce a reduction in cutting forces and machining temperature. Compared to that, changes in relief angle showed little effect [14, 15].

The effects of rounding of the cutting edge and width of the cutting tool margin complement each other. The effect of margin width is approximately linear: A larger margin width increases the contact area at the tool–workpiece interface, leading to greater frictional force, instability of machining process, and larger cutting forces. In contrast, changes in the radius of the cutting edge have nonlinear effects and depend on the margin width: For a range of rounding radii between 9 and 25 μm, the change in the cutting force follows either a concave down or a concave up curve when the margin width is below or above 80 μm, respectively. In all of the cases, the minimum or the maximum cutting forces occur around a rounding radius of 17–18 μm. Therefore, the smallest cutting forces occur when the rounding radius is around 17–18 μm, and the lowest possible margin width is used [15, 20].

Chip formation and morphology are another important area in HSM as they are primarily dependent on the cutting speed, which is the largest contributor to HSM conditions. A recent development in studies regarding chip formation has been the utilization of dislocation and plasticity theories in predicting and simulating this process in HSM. As the material is being machined, it experiences strong plastic deformation leading to dislocation migration and accumulation at the shear area near the cutting point with periodic distribution of said dislocations. As the cutting speed increases moving from conventional to HSM area, the strain and strain rate of the material enlarge the effects of dislocation accumulation, which leads to a more serrated shape and higher degree of segmentation of the formed chips, as can be seen in Fig. 2.4. Xiuli et al. [21] applied these theories in numerical simulations of HSM of AA7050. It was found that a cutting speed below 1500 m/min led to the formation of a continuous ribbon chip; when the speed is increased above that point, the chip tends to form more serrated, and it finally fully becomes serrated with a high serration degree when the cutting speed goes beyond 2500 m/min [21].

In recent years, there have also been studies in numerical simulations of aluminum alloys. Wang et al. [22] proposed a modified version of Johnson–Cook constitutive model to predict the behavior of AA7050 during HSM. The modified model accounts for strain and strain rate depending on the temperature, which is not the

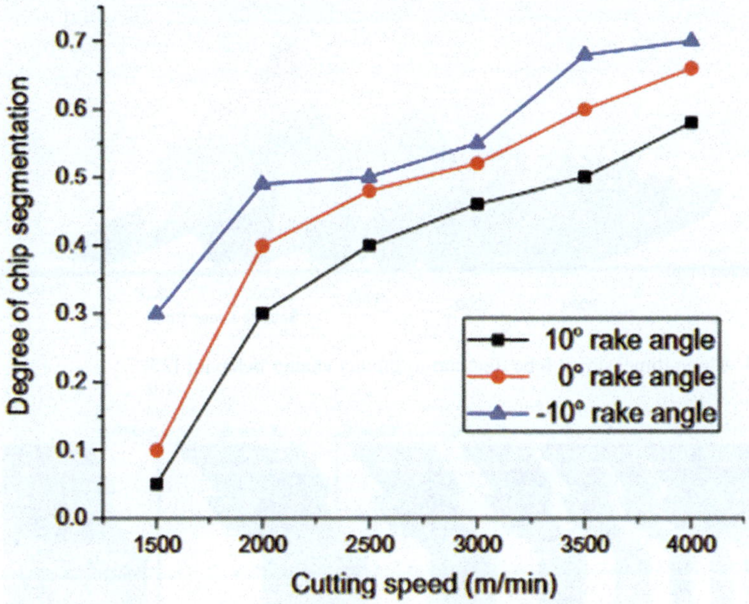

Fig. 2.4 Effects of cutting speed and rake angle on chip segmentation [21]

case for typical Johnson–Cook models. Such changes allowed to reduce the prediction error from 6.89% to 3.79% for average cutting force, from 20.26% to 12.03% for average thrust force, from 15.64% to 9.18% for chip height, and from 12.20% to 6.22% for the height of continuous sections.

Another approach to the improvement of numerical simulations was the choice and balancing of the input parameters for Johnson–Cook model. Wang et al. [23] looked into five different parametric inputs from the existing studies on the material properties of AA7075 and compared their predictions with experimental values. The JC5 model from a study by Paturi et al. [24] showed the most accurate results: The error in the horizontal forces was 5.73% and 3.13% for x- and y-components, and the error in the vertical force is 10.69%.

On the other hand, chatter is a cause of many negative effects during machining and leads to a reduction in tool life and surface quality. It is more prominent in the machining of thin-walled parts, which is a common area of HSM applications for aluminum alloys. To combat that, in machining of AA7075, a three-dimensional lobe diagram was developed by Damous et al. [25]. The lobe diagram, as seen in Fig. 2.5, depends on tool position, spindle speed, and the depth of cut. From the diagram, it can be seen that one should stay within the "islands" underneath the curve by modifying those parameters. Higher spindle speed brings about larger stability islands, whereas lower depth of cut leads to increasing the size of those islands. Figure 2.6 shows an example of a stability map obtained from the lobe diagram for a 3-mm depth of cut.

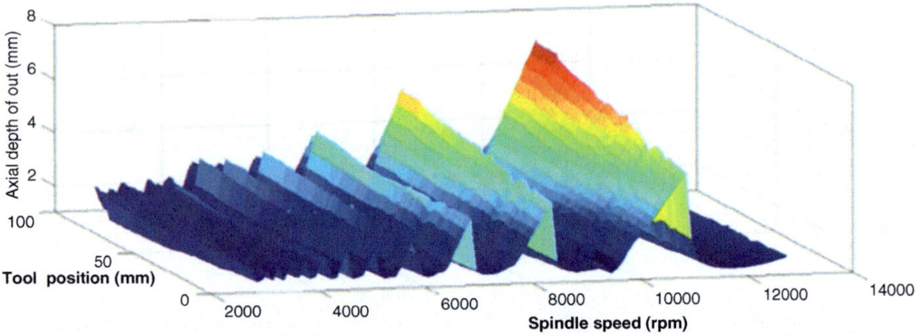

Fig. 2.5 Three-dimensional lobe diagram to predict chatter behavior [25]

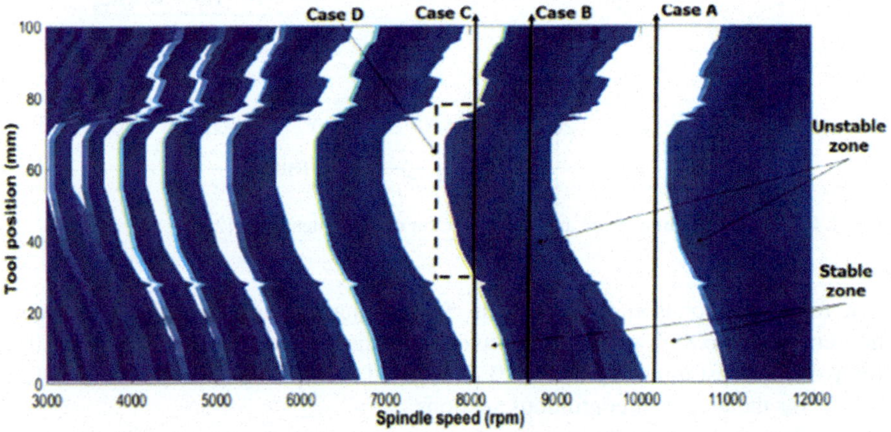

Fig. 2.6 Two-dimensional stability map for the depth of cut of 3 mm [25]

Another recent development in HSM of aluminum alloys was the introduction of water jet peening (WJP) composite surface reinforcement during the machining process to reduce corrosion of AA7075. This process noticeably improves corrosion resistance and decreases the size of grains. Water jet pressure, trajectory spacing, and nozzle speed have various effects on the resulting corrosion resistance: with an increase in water jet pressure, polarization resistance and self-corrosion potential decrease while self-corrosion current first increases and then decreases; increasing the spacing initially leads to the growth of potential and current density as well as polarization resistance but further increases correspond to reduction in all of them; and lastly, increasing the nozzle speed brings about positive effects on the surface quality of the material [26].

Recently, the degree of precompression deformation of AA7050 has been investigated with regard to its effects on surface quality and work hardening during HSM at various parameter settings. Overall, from the tested 10%, 15%, and 20% predeformation levels, 15% showed the lowest surface roughness and the best

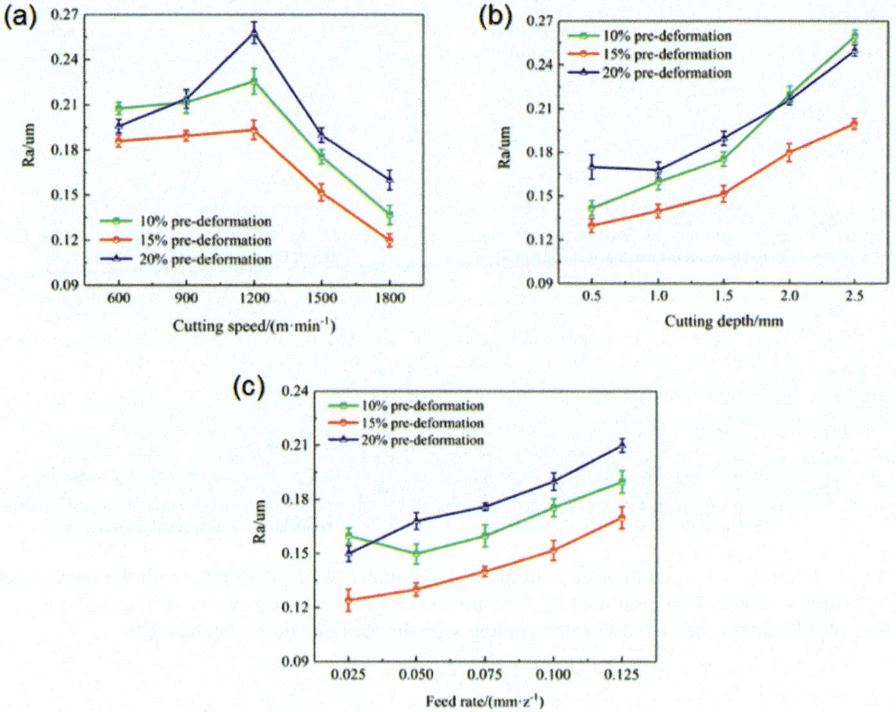

Fig. 2.7 Effects of HSM parameters on surface roughness at different predeformation levels: (**a**) cutting speed; (**b**) cutting depth; (**c**) feed rate [27]

surface quality, while 20% showed the highest roughness and the lowest quality. Figure 2.7 shows the effects of HSM parameter choice on different crystal orientations with predeformation. In addition, work hardening was shown to be the highest at 20% predeformation, followed by 10% and 15%. Overall, work hardening shows initial growth followed by a decrease when the cutting speed and depth of cut are increased, while increasing the feed rate positively affects its layer depth [27].

2.2.2 Turning of Aluminum Alloys

HSM has an advantage over conventional machining in its capability to conduct the machining operations in dry conditions because most of the heat is removed with the chips. However, cooling is still an important consideration during the process. The cooling solutions are similar in turning and milling. In recent years, the application of minimum quantity lubrication (MQL) and cryogenic cooling with liquid nitrogen has been tested in high-speed turning of AA7075 alloy against dry conditions. MQL and cryogenic cooling successfully reduced the temperature with

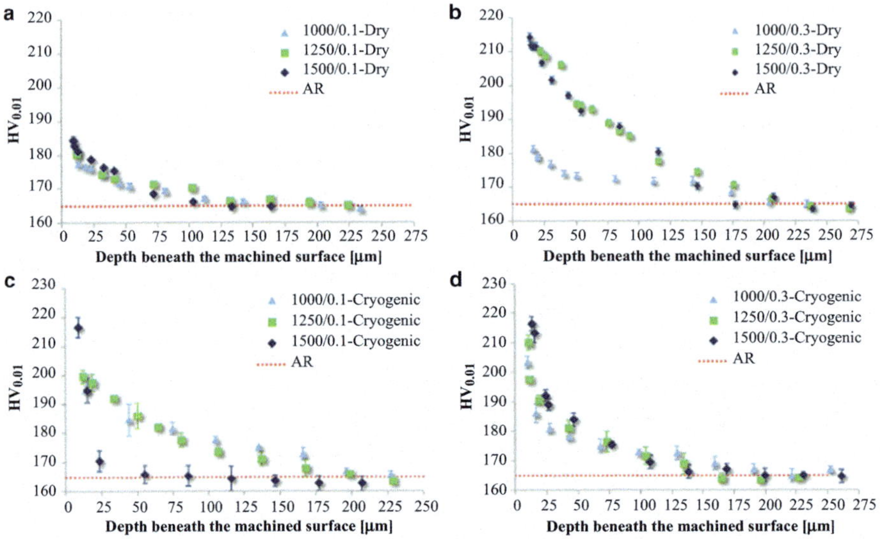

Fig. 2.8 Changes in micro-hardness in the material under (**a**) dry condition with the feed rate of 0.1 mm/rev, (**b**) dry condition with the feed rate of 0.3 mm/rev, (**c**) cryogenic cooling with the feed rate of 0.1 mm/rev, and (**d**) cryogenic cooling with the feed rate of 0.3 mm/rev [29]

cryogenic cooling showing more significant results. In addition, these cooling methods reduce rake and flank wear of the tools with cryogenic cooling showing better results on average, especially at higher feed rates. Overall, the tool wear was reduced and more affected by cooling on the rake face [28].

Cryogenic fluid application was further explored in the study by Imbrogno et al. [29]. Cryogenic cooling contributes to a reduction in the grain size and improves creep pitting corrosion resistance by decreasing the number of copper particles underneath the surface layer but increases the surface roughness and tool wear on the flank face by embrittling the tool and the material. In addition to that, cryogenic fluid cooling leads to higher hardness in the resulting product at a smaller feed rate due to precipitation strengthening and finer grains. Dry cutting conditions lead to smaller hardness mostly due to precipitate dissolution caused by higher temperatures during the process. Figure 2.8 shows the exact comparison of both cooling conditions at various cutting speeds and feed rates.

2.3 Machining of Titanium Alloys

Titanium alloys, specifically Ti-6Al-4V, are ubiquitously used in the aerospace industry for critical parts of aircraft. For example, they are used in turbine-engine fan blades, the disks to which the fan blades are attached, or in low-pressure compressors [1]. This alloy possesses a number of highly beneficial properties to aerospace applications: high strength-to-weight ratio, high corrosion, creep and fatigue resistance, and great performance at high temperatures due to retention of its mechanical properties [9].

Ti-6Al-4V is considered to be a difficult-to-cut material due to several issues that arise from its thermal and mechanical properties during machining: high heat generation, tool wear, large cutting forces, build-up edge, tool chatter, and breakage [9, 30]. As the effort to higher machining efficiency and productivity rises, HSM becomes more prevalent in the aerospace industry giving a higher need to combat the aforementioned difficulties. Therefore, multiple studies have been conducted in recent years focusing on the reduction of tool wear, improvements to cooling, prediction of chip behavior, and development of new technologies.

2.3.1 Milling of Titanium Alloys

As Ti-6Al-4V is a difficult-to-cut material, it has significant effects on the tool life during HSM. Increasing the cutting speed leads to both a reduction in the machining time and a reduction in the tool life. Therefore, there should be a balance in order to benefit from both. For uncoated carbide tools, the best ratio of tool life compared with machining time is at a cutting speed of 130 m/min according to Matuszak et al. [31]. Further increasing the speed up to 190 m/min reduces the surface roughness of the machined part followed by an increase at higher speeds. In addition to that, 190 m/min results in the largest fatigue resistance, the deepest hardened layer of 28 μm, and the highest residual compressive stresses. However, increasing the speed past that significantly increases the tool wear with a sharp rise in flank wear. The improved hardness has also been observed by Ullah et al. [32], who further investigated the effects of various parameters on the surface layer. Figure 2.9 shows the effects of cutting speed and feed rate on surface hardness.

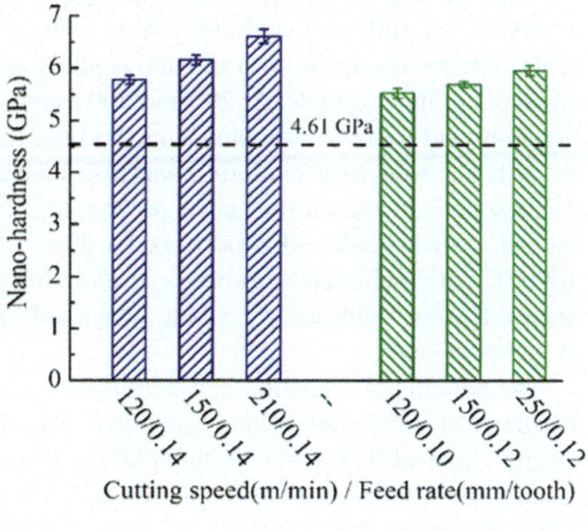

Fig. 2.9 Surface hardness based on cutting speed and feed rate [32]

Understanding chip formation during high-speed milling of Ti-6Al-4V provides a significant insight into the machining process and tool wear. During this process, cutting speed plays a significant role in chip morphology. Initially, the cutting forces increase with the increase in cutting speed and the chips are continuous. Between 125 and 250 m/min, chip serration begins and is increased, whereas the fracture type changes from shear to tensile. Further increase in the cutting speed results in shear fracture being able to overcome tensile fracture, which corresponds to the reduction of cutting forces and degree of serration. The processes and their effects are connected to the nanograin structure: At around 250 m/min, they have columnar shape due to tensile fracture type; when the speed reaches 375–500 m/min, they recrystallize and appear equiaxed due to thermal softening [33].

Similarly, Ullah et al. [34] numerically investigated the serrated chip formation during high-speed milling of Ti-6Al-4V using speeds up to 250 m/min further going into the formation of serrated chips. Serrated chip formation occurred due to shear localization propagating from the free surface of the workpiece far toward the primary shear zone. With the increase in cutting speed, the temperature of the chip increases, while the back side of the chip goes through strong plastic deformation with no phase transition.

High-speed milling of a difficult-to-cut material, such as Ti-6Al-4V, has significant effects on the tool life. In order to prolong it under such conditions, various coatings can be applied. Monolayer TiCN, AlTiN, and TiAlN, as well as two-layer TiCN + TiN and AlTiN + TiN, were investigated by Kuram [35] in HSM at a cutting speed of 100 m/min. During machining, all of these coatings result in the same chip morphology and evolution. Moreover, coating delamination and adhesion wear are also found in all of these coatings, whereas crater wear appears on all of them but TiCN coating, and rake face cracking is detected only on AlTiN coating. In addition, TiCN coating results in the production of the smallest surface roughness and cutting forces at the end of machining, whereas TiAlN does that at the start. Considering tool wear, the study concluded the following: TiCN results in the lowest wear followed by AlTiN + TiN, TiCN + TiN, TiAlN, and AlTiN in the increasing order.

Wang et al. [36] further investigated the tool wear mechanisms when machining Ti-6Al-4V at 100, 200, and 300 m/min using PVD-TiAlN-coated tungsten carbide tools. The results agree with the above study at 100 m/min; however, the behavior changes at higher speeds. At 200 and 300 m/min, chipping, flaking, and cracking are the dominant tool wear mechanisms, and the tool wear was significantly increased overall. The tool wear evolution was also investigated: First, the coating delamination happens at the cutting edge propagating to the rake face and the flank face; second, cobalt binder diffusion occurs due to high temperatures reaching around 1000 °C with higher cutting speeds accelerating the process; and finally, microcracks appeared at the voids left by cobalt, which led to crack initiation and brittle fractures at higher speeds.

The benefits of a coating can be further enhanced by the application of cooling. Khatri et al. [37] studied and compared dry, flood cooling, and MQL conditions during HSM of Ti-6Al-4V with TiAlN-coated carbide tool. Among these cooling conditions, MQL results in the lowest amount of abrasion, adhesion, and chipping wear. Therefore, it is a valuable additional method to improve the tool life.

As cooling further increases, the tool life when high-speed milling a difficult-to-cut alloy such as Ti-6Al-4V, various methods are explored and weighed against each other. During HSM in dry, flood-coolant, and MQL conditions, abrasion wear on the rake and flank faces is the most prominent tool wear mechanism followed by chipping and adhesion wear. Flood coolant shows the largest reduction in abrasion wear, followed by MQL. However, MQL outperforms it when it comes to chipping and adhesion wear. MQL additionally improves the surface roughness and promotes continuous chip formation in addition to overall being a more sustainable cooling option [38].

MQL can be further improved by the addition of supercritical CO_2. When compared to flood coolant, it showed the following improvements: a reduction in chip width and burr's height, an increase in tool life by four times in short processes and by seven times in long processes, a reduction in cutting forces by 31% for short processes and 50% for long processes, an increase in microhardness by 30% for both the processes, a decrease in surface roughness by up to 43% for short processes and by 35% for long processes, and, lastly, a reduction in plastic deformation resulting in lower cutting temperature. It was estimated by Khosravi et al. [39] that the profit margin can be improved by around two times when used for Ti-6Al-4V in HSM.

The application of ultrasonic vibration is highly beneficial in HSM of Ti-6Al-4V. One of the ways to apply vibration to a milling process, longitudinal–torsional ultrasonic vibration milling (LTUVM), was investigated by Han et al. [40]. Despite small effects on milling forces in HSM compared with its application in conventional milling (CM), it has significant benefits in tool wear reduction, which is amplified with the rise in MRR, and Fig. 2.10 shows this effect.

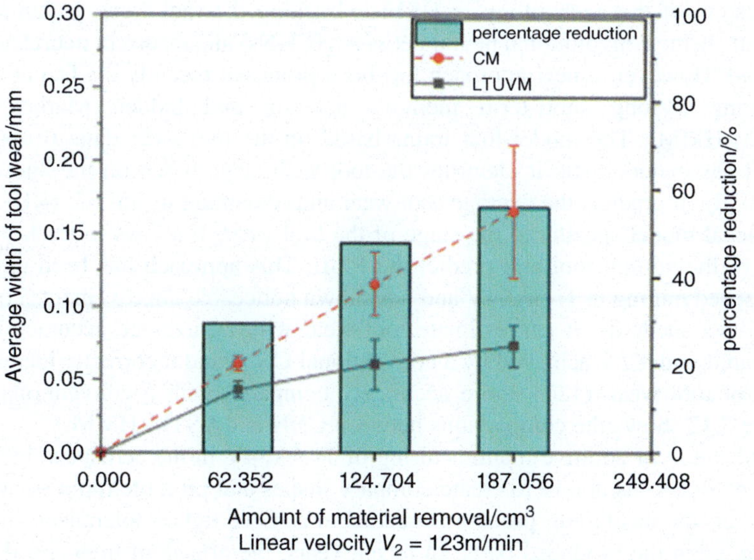

Fig. 2.10 Effects of LTUVM on tool wear compared with conventional milling (CM) [40]

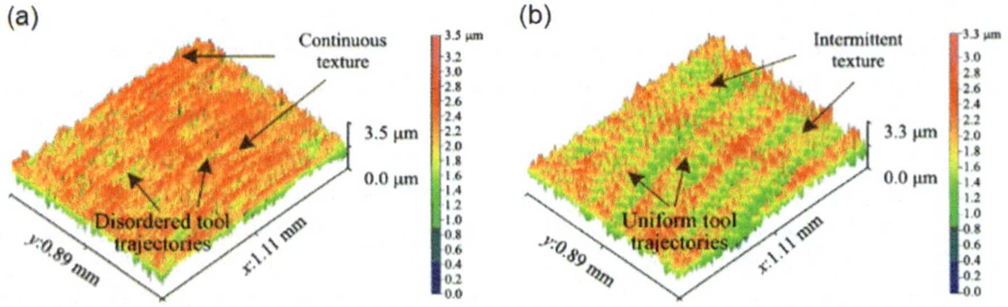

Fig. 2.11 Surface morphology of machined part: (**a**) conventional milling, (**b**) UVAM [41]

A new two-dimensional (2D) ultrasonic vibration-assisted milling (UVAM) platform has been built and studied by Zhang et al. [41]. At its optimal parameters, the platform is capable of reducing temperature by 70.68%, improving surface quality, and increasing the maximum value of residual compressive stresses. Figure 2.11 shows the comparison of surface morphology between conventional milling and UVAM.

Last, high-pressure coolant can also be applied alongside ultrasonic vibration to improve high-speed milling of Ti-6Al-4V. At the pressure of 200 bar and a cutting speed of 160 m/min, the tool life is increased by 3.4 times, the maximum MRR is increased by 135%, and temperature is decreased by around 20%. The application of high-pressure coolant provides a better cooling effect with an addition of vibration, therefore significantly reducing adhesive wear [42].

Advances in machine learning and neural networks in recent years allow to achieve a higher level of understanding of the machining process. An important application of this technology in HSM is tracking the tool wear. It can be done typically using convolutional neural networks (CNNs) and recurrent neural networks (RNNs). However, a new approach has been proposed recently by Tao et al. [43], involving a long short-term memory network and hidden Markov model (LSTM-HMM). The model first trains based on the tool wear data to update the model parameters; then, it identifies the tool wear stage based on the known data, and, lastly, it predicts the average tool wear and remaining useful life (RUL) based on a local model specific to the stage of the tool wear. It allows to track the flank wear of the cutting tool and predict the RUL. This approach has been applied to high-speed milling in Ti-6Al-4V and has shown noticeable improvements compared with other methods. It can estimate tool wear with an average accuracy of 95% compared with 87% achieved by a conventional CNN, and it correctly identifies the stage of tool wear 11.7% more accurately compared with a conventional RNN. Figure 2.12 shows the comparisons between CNN and LSTM-HMM.

Utilization of additive manufacturing of Ti-6Al-4V in the aerospace industry is growing due to its ability to create complex shapes and promote rapid prototyping. However, the machining process is needed to achieve tighter tolerances and better accuracies in the produced parts. In recent years, the effects of high-speed milling using a ceramic tool have been investigated on such parts. At speeds between

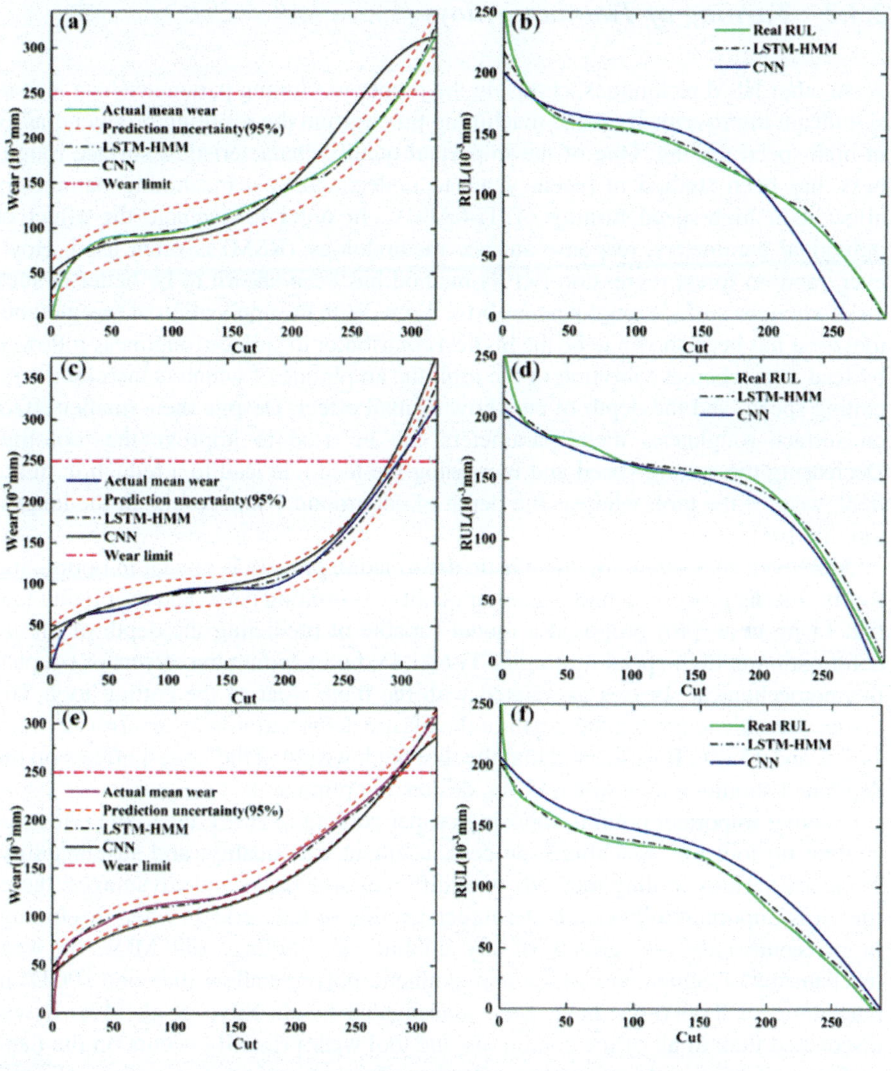

Fig. 2.12 Comparison of CNN and LSTM-HMM in terms of tool wear and RUL [43]

150 and 250 m/min, feed rate shows a higher effect on the cutting forces compared with the cutting speed. Adhesion, abrasion, and microchipping are observed to affect the tool wear: adhesion primarily on the rake face and abrasion and adhesion together on the flank face. The chips produced during the process showed signs of serration on the free surface and were continuous with occasional breakage. The average surface roughness achieved by this process is within 2 μm and depends on the cutting speed and the feed rate [44].

2.3.2 Turning of Titanium Alloys

As in other HSM techniques, choosing the proper machining parameters can lead to significant improvements in the machining process and the resulting product quality in high-speed turning. One of the important quality characteristics, surface roughness, has been studied in recent years to understand how machining parameters affect it in high-speed turning of Ti-6Al-4V. In order to evaluate the effect of individual parameters, response surface methodology (RSM) is often used. However, random forest regression (RFR) method has been shown to be better at such tasks when the surface roughness is above 1 μm. With the application of this method, tool wear has been shown to be the highest contributor to surface roughness followed by feed rate, whereas vibration of the tool, the amplitude of which is increased with cutting speed, and the depth of cut show a small effect. Despite their smaller effect on surface roughness, these parameters can be used to improve the tool life: Decreasing the cutting speed and increasing the feed rate lead to a reduction in the flank wear of the tool, whereas the depth of cut around 1 mm results in the longest tool life [45].

Moreover, understanding the plastic deformation process is crucial in optimizing the machining process, achieving better quality of resulting parts, and improving tool life. Liang et al. [46] proposed a model capable of predicting the depth of plastic deformation in high-speed turning of Ti-6Al-4V by including the effects of coupled thermomechanical stresses associated with the flank wear of the cutting tool. The model was able to predict the depth of the plastic deformation with an error between 11.2% and 15.4%. It was found that the tool flank wear had the largest effects on the depth and should be controlled if one desires to optimize it.

Another important point is choosing proper cooling in HSM due to the effects of cooling on tool life, machining process, resultant part quality, and sustainability. There are various cooling methods with different drawbacks and advantages; therefore, it is important to have a better understanding of their effects when machining. In a recent study, the effects of dry cooling, jet cooling, and MQL on wear mechanisms of cemented carbide and synthetic polycrystalline diamond (PCD) in machining of Ti-6Al-4V have been investigated by da Silva et al. [47]. It was discovered that, in all of the conditions, the tool wear primarily occurs on the flank surface, and it is mainly affected by adhesion, abrasion, and attrition sources of wear. For all of the cooling conditions, a smaller feed rate results in a longer tool life. Jet cooling conditions lead to the longest tool life, followed by MQL and dry conditions. Under jet cooling, PCD shows a significantly longer tool life compared with cemented carbide. However, under dry and MQL conditions, cemented carbide shows comparable results and may be used as a cheaper alternative. Figure 2.13 shows a more precise comparison of tool wear under different cooling conditions.

Several coating types and methods have been studied in recent years. AlTiN and CrN hard PVD coatings are some of the often used ones. The comparison of these coatings has shown that CrN outperforms AlTiN in high-speed turning of Ti-6Al-4V when used on a cemented carbide tool. CrN significantly increases the tool life by

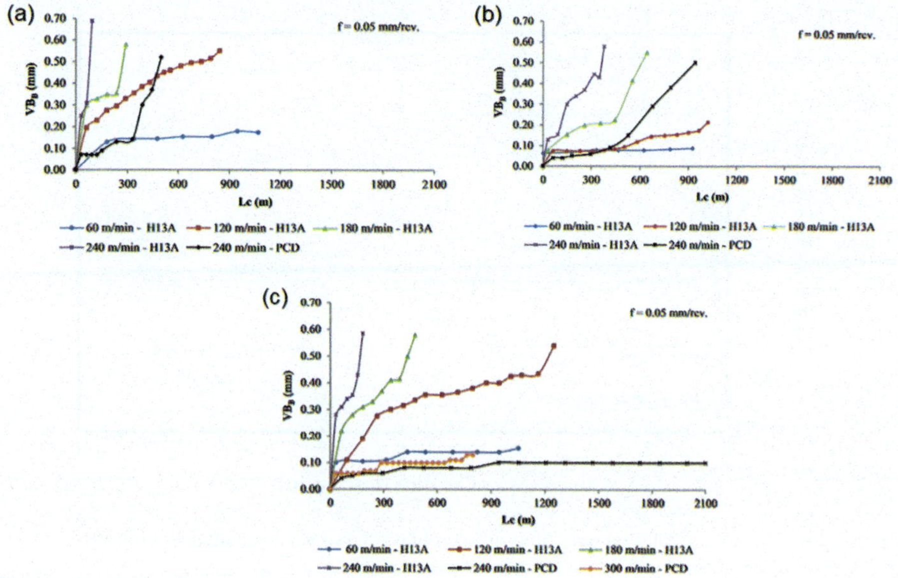

Fig. 2.13 PCD and cemented carbide flank wear—comparison at a feed rate of 0.05 mm/rev under (**a**) dry, (**b**) MQL, and (**c**) jet cooling conditions [47]

reducing the crater wear due to the formation of thermal barrier Cr_2O_3 tribofilms during operation. AlTiN, in comparison, performs better than an uncoated tool in terms of built-up volume, but it shows increased crater wear [48]. Nevertheless, the performance of PVD-TiAlN can be improved when microblasting is added to that. Microblasting lowers surface roughness and improves tool morphology, which results in better surface hardness and residual compressive stresses. After this operation, the tool life is increased by 45% compared with an untreated tool. Treated tool delays the evolution of the tool wear during its initial, early stable, and middle stable stages. However, when the tool wear enters the late stable and rapid stages, it shows a sharp increase, leading up to the end of the tool life [49], and Fig. 2.14 shows this behavior.

2.3.3 Grinding of Titanium Alloys

In recent years, intermittent feed high-speed grinding was investigated as a potential way to reduce heat generation during the process. Yi et al. [50] studied its application in machining of Ti-6Al-4V. This process involves a periodical retreat of the grinding wheel to allow the grinding fluid to be supplied directly into the working zone. Such process leads to a significant reduction in temperature, reduces the abrasive wear of the grinding wheel, and improves the surface integrity of the resulting workpiece.

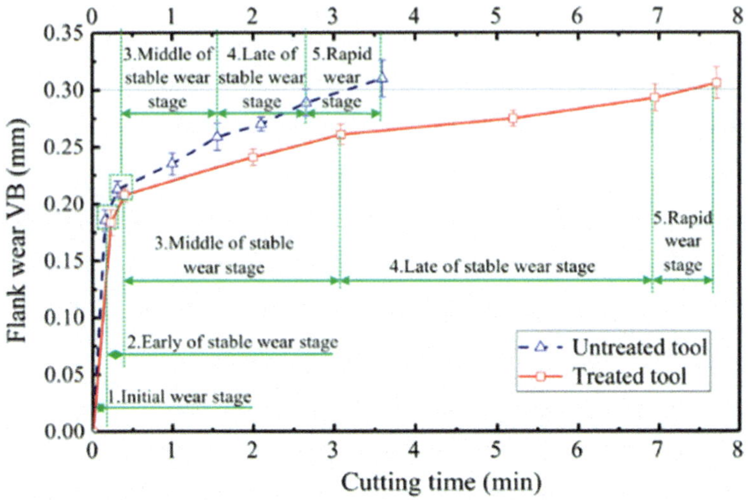

Fig. 2.14 Tool wear evolution over time of microblasted tool versus untreated tool [49]

However, the main drawbacks of such process are a significant increase in machining time and a reduction in efficiency compared with continuous high-speed grinding.

Ultrasonic vibration and MQL have different positive effects on the high-speed grinding process. However, when it comes to combining them in a single process, it is important to understand the effects they have on each other. In a recent study by Naskar et al. [51], ultrasonic-assisted grinding was applied alongside MQL to see such effects. It was found that the choice of liquid for MQL had significant effects on the vibration in ultrasonic-assisted grinding. Liquids with higher lubricity provide better transfer of vibrations, leading to higher vibration amplitude. Consequently, it results in increased height and pitch of the formed scallops, therefore increasing surface roughness and improving the residual stresses.

2.3.4 *Drilling of Titanium Alloys*

Choosing the proper parameters in high-speed drilling has a significant effect on the productivity of the process. The effects of the spindle speed, feed rate, and choice of the tool material have been investigated in a recent study [52] with regard to drilled hole circularity, flank tool wear, and the temperature in dry drilling of Ti-6Al-4V. The following variation in parameters was present: spindle speed between 4200 and 10,600 rpm; feed rate between 159 and 400 mm/min; and choice of material between high-speed steel, uncoated carbide, and TiN-coated carbide. It was found that drill bit material had the highest effect on the process followed by rotational speed and

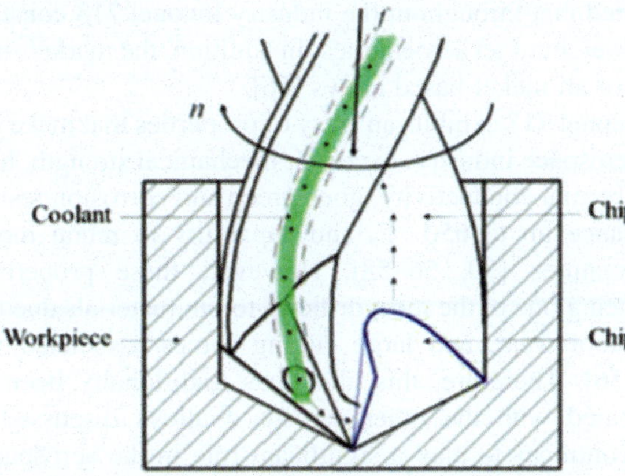

Fig. 2.15 Diagram of internal cooling in a twist drill [54]

feed rate. The optimal parameters within the investigated range are the following: tungsten carbide twist drill coated with TiN, a rotational speed of 10,600 rpm, and a feed rate of 400 mm/min.

In high-speed drilling, similar to the other processes, application of a coolant is highly beneficial to the machining process. According to Hari Chealvan et al. [53], cryogenic cooling offers advantages over wet and MQL cooling in its removal and disposal. Application of high-pressure cryogenic coolant at 22 bar shows a significant reduction in temperature, thrust force, and torque. Increasing the pressure of the said coolant leads to better penetration at the tool–chip interface and therefore takes the role of a lubricant in reducing frictional forces and improving the quality of produced holes.

Another way to improve the high-speed drilling process was developed by Guo et al. [54]. They proposed two drilling tool designs: carbide internal cooling twist with microgroove textures on rake face or flank face. The microgrooves, which are produced by neodymium-doped yttrium aluminum garnet (Nd:YAG) laser, showed a significant improvement in the machining process depending on the location of the microgrooves. Grooves on the rake face improved chip-curling, reduced the contact length between the tool and the chips, and improved chip breakage. In comparison, grooves on the flank face significantly improved internal cooling, which led to a reduction in cutting forces and a decrease in abrasive and adhesive wear. Figure 2.15 shows the diagram of the studied drilling tool.

2.4 Machining of Inconel

Nickel-based alloys, such as Inconel 718, are another family of alloys that are widespread across the aerospace industry due to their sought-after properties. According to Boyer et al. [1], aeroengines produced by General Electric Aviation in 2001 comprised of around 35% Inconel 718 in terms of weight. Ulutan et al. [55]

reported that throughout the industry Inconel 718 constituted more than 50% of the total weight of jet aeroengines. In addition, the market share of Inconel 718 is around 54% of all nickel-based alloys [56].

Inconel 718 exhibits an array of properties that make it particularly well-suited for the aerospace industry: excellent mechanical strength, high strength-to-weight ratio, low thermal conductivity, good creep and corrosion resistance, great thermal fatigue resistance up to 650 °C, and capability to retain mechanical properties at high temperatures [30, 56–58]. However, these properties also result in Inconel 718 being one of the most difficult-to-cut materials due to excessive heat generation, high tool wear, and large cutting forces associated with its machining process [30, 56]. Therefore, this alloy has traditionally been machined at lower speeds compared with the other aerospace alloys discussed in this work. As Inconel 718 continues to play a significant role in the aerospace industry, the demand for higher productivity and efficiency is on the rise. The advancements in HSM play an important role in meeting such demand. In this section, the recent progress in cooling, prevention of tool wear, choice of tools and their coatings, improvements in the surface quality, predictions of the machining behavior, and introduction of new machining techniques and tools are discussed.

2.4.1 Milling of Inconel

Investigating the tool wear mechanism and what it is affected by is an important point of studies in HSM of a difficult-to-cut alloy, such as Inconel 718. In recent years, it was studied with a new type of approach—the analysis of individual flutes on the tool. Gueli et al. [58], who investigated it on an uncoated tungsten carbide tool at 60 m/min, found that increasing both the depth of cut and the feed rate led to higher tool wear. The wear is more prevalent on the rake and back areas of the flutes, and burrs form at the depth of cut larger than 0.5 mm. In addition, surface roughness is lower at the entrance compared with the exit of the machined area.

Sometimes, it is necessary to machine a material with a curved surface. A ball-end milling tool with AlTiN could be used for that in HSM. The curved shape of the surface affects the machining process by making the cutting forces larger and more volatile, which also increases the tool wear. Moreover, tool wear is different when machining a curved surface compared with a flat surface: Curvature leads to rake wear being the main type of tool wear followed by flank wear and chipping compared with flank wear and coating damage being the most prominent on a flat surface. In order to increase the tool life, one could change several characteristics of the tool geometry: decrease the rake angle to achieve smaller cutting forces due to increased chip–tool contact area, increase the relief angle to reduce flank wear and cutting forces, and avoid helix angles above 30° [57].

Chip formation, as discussed earlier, is an important process in tracking machining characteristics and behavior. Recently, there have been advancements in analyzing and predicting it in HSM of Inconel 718 using finite element analysis (FEA)

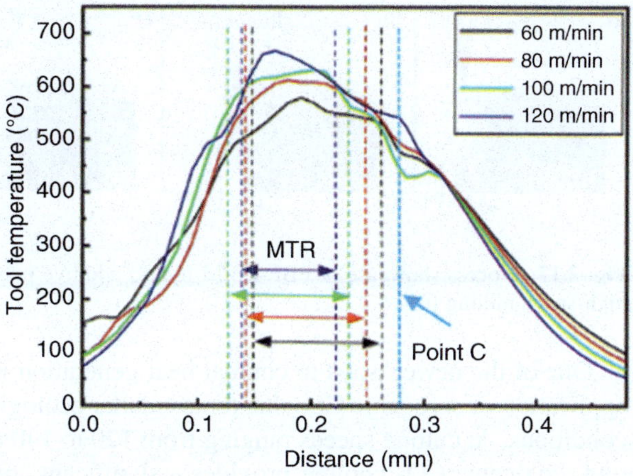

Fig. 2.16 Temperature variation along with the surface of TiAlN-coated tool. The left border is the top of the rake face, point C is the cutting edge, and the right border is the top of the flank face [59]

method. As cutting speeds rise toward 120 m/min, the maximum temperature region moves away from the tool tip toward the top of the rake face (as can be seen in Fig. 2.16) reaching temperatures above 660 °C. At the same time, sawtooth chip formation degree increases, leading to the growth of the amplitude of cutting force fluctuation. Such behavior leads to higher tool wear on the rake face, resulting in decreased hardness and strength of the cutting edge, which further increases the damage in the machining process [59].

A recent advancement in numerical simulation of HSM of Inconel 718 was incorporating Hall–Petch hardness prediction model and dynamic interaction between mobile and immobile dislocations into FEA simulation. Rinaldi et al. [60], having developed this model, were able to predict cutting forces with 4% total absolute error, hardness with 3% mean absolute error, grain size with under 3% absolute error, and chip morphology with 11% maximum absolute error.

As Inconel 718 is a very difficult-to-cut material, the tool wear even in carbide tools is significant. Recent studies in tool choice have investigated the usage of full-body ceramic tools. Tested at a cutting speed of 600 m/min, they were compared to conventional milling with carbide tools at 40 m/min at several cooling conditions. Under dry milling conditions, ceramic tools had 35% longer tool life, and with additional air-blasting helping with chip evacuation, it increased to 39%. Ceramic tools were used at a cutting speed of 600 m/min, which corresponded to HSM conditions, while carbide tool was used at a cutting speed of 40 m/min, which is conventional milling. The dominant tool wear was found to be chipping of the tool wear that was significantly higher in MQL conditions due to fast temperature changes. The authors recommended avoiding MQL with ceramic tools. Ceramic tools also show significant improvements in material removal over their lifetime when used in HSM: Air-blast-assisted dry milling with ceramic tool was 2.37 better than dry milling with carbide tool, 1.82 better than carbide with MQL, and 1.17 better than carbide with flood coolant [61].

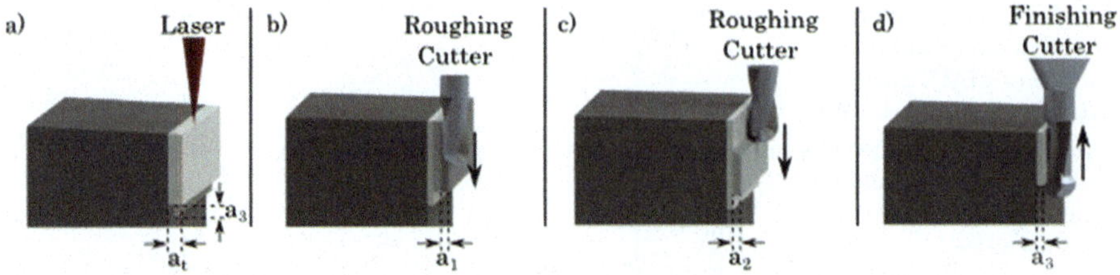

Fig. 2.17 Process showcase of PBF-LB/M hybrid additive manufacturing integrated with in situ high-speed milling [63]

One of the newer ways to combat heat generation in HSM of Inconel 718 is the application of cryogenic CO_2 due to its smaller ecological impact and safer working conditions. At cutting speeds ranging from 120 to 140 m/min using a coated carbide tool, cryogenic CO_2 cooling provides a significant consistent positive effect at the cutting area by decreasing the cutting temperature by 80% and increasing the tool life by 70.8% compared with dry cutting. The tool wear evolution under cryogenic CO_2 conditions is similar to dry conditions: Smooth abrasion and chipping begin close to the cutting edge followed by flank wear and finally resulting in flaking at the rake face and notching at the flank face [62].

With the advances in metal additive manufacturing, powder bed fusion – laser beam melting (PFD-LB/M) allows the production of complex geometries and faster prototypes. However, accuracy and surface quality are still considered challenges yet to be resolved. Therefore, a hybrid approach utilizing HSM methods could be used to combat these challenges. Integrating high-speed milling directly into the additive manufacturing chamber allows for an even faster manufacturing process, which is showcased in Fig. 2.17.

Choosing optimal parameters in high-speed milling creates many benefits including increased surface quality. Sommer et al. [63] have investigated optimization of machining parameters for this process when manufacturing Inconel 718. The following optimization parameters are discussed in order of importance. The main parameter affecting the process is infeed that, when decreased, reduces the flank tool wear and reduces surface roughness. It is followed by spindle speed that can lead to cracks in the flank face at lower values and results in heat damage due to higher temperatures at higher speeds; both of them affect the surface roughness. Next is z-pitch that, when decreased, leads to a small reduction in surface roughness due to smaller waviness at the surface profile; however, it increases the machining time. Lastly, feed rate and consequently feed rate per tooth affected the surface roughness the least; however, the actual effect of these parameters would increase or decrease depending on whether the spindle speed is smaller or larger, respectively.

The optimal parameters that led to the lowest surface roughness according to Sommer et al. [63] achieved the following results: average surface roughness $R_a = 0.47$ μm and maximum surface roughness $R_z = 3.73$ μm. The following parameters led to these results: an infeed of 10 μm, a spindle speed of 9600 rpm, a z-pitch of 80 μm, and a feed rate of 240 mm/min.

2.4.2 Turning of Inconel

In high-speed turning, cubic boron nitride (CBN) and polycrystalline cubic boron nitride (PCBN) tools have attained recent interest in applications in high-speed turning of Inconel 718 due to their high toughness, ability to retain hardness at high temperatures, sharp cutting edges, and capability to be used in high-speed applications. Various studies have been conducted investigating the effects of the application of such tools, wear mechanisms, and comparison with the commonly used tools.

Tool inserts with 65% CBN content bound by TiAlN were recently investigated by Tu et al. [64]. It was found that the main sources of tool wear were adhesive wear on rake and flank faces and cutting-edge chipping due to periodically changing strain hardening and thermal softening processes. However, as the cutting speed became larger, thermal softening became more prevalent compared with strain hardening. Furthermore, the compressive stresses appearing in the workpiece during the process tend to increase with the flank wear and decrease with the cutting speed. Compared to that, PCBN tool wear evolution was the following: chipping, notch wear, flank wear, and built-up edge in the beginning; after around 5 min, crater wear started occurring, and lastly, flank wear became the dominant type of wear [65]. Another study claimed the wear mechanism to be the following: notching wear being the most prevalent followed by adhesion, chipping, and crater wear in no specific order [66]. Overall, these findings complement each other with the first one going more in-depth in how this process changes over time and the second one describing the mechanisms overall.

The tool life is primarily affected by increases in the cutting speed and CBN content with feed rate increases showing an insignificant effect. The longest tool life was achieved at a cutting speed of 250 m/min, a feed rate of 0.15 mm/rev, and a CBN content of 45%. In dry high-speed turning conditions, the tool life of PCBN is below 2 min, whereas, with the addition of fluid coolant, it extends to 5–10 min depending on the tool. Therefore, dry high-speed turning was deemed not viable for Inconel 718 by the authors. In comparison, carbide tools' life is around 30 min under dry conditions and above 40 min with the coolant. Despite the shorter tool life, the PCBN tools have the following benefits compared with carbide tools: significant reduction in surface roughness (0.2–0.8 μm versus 2 μm) and increase of machined surface per cutting edge of up to 155% [65, 66].

Tool coating is another way to improve tool life and surface quality of a material without changing the tool itself. Positive TiAlN, negative TiAlN, and negative (Al, Ti)N are some of the coatings that could be used in high-speed turning of Inconel 718. These coatings have been investigated by Zhang et al. [67] at cutting speeds of 120–200 m/min. On the skirt flank, the primary tool wear mechanisms are adhesive and abrasive for all of the coatings, while rake wear is the highest on positive TiAlN coating, followed by negative TiAlN coating, with (Al, Ti)N coating affected the least. The cutting-edge flank wear mechanisms differ for all the coatings, with (Al, Ti)N being affected by chipping and built-up edge at higher speeds. Overall, the negative (Al, Ti)N coating led to the highest tool life and resulting hardness of the machined part as can be seen in Figs. 2.18 and 2.19.

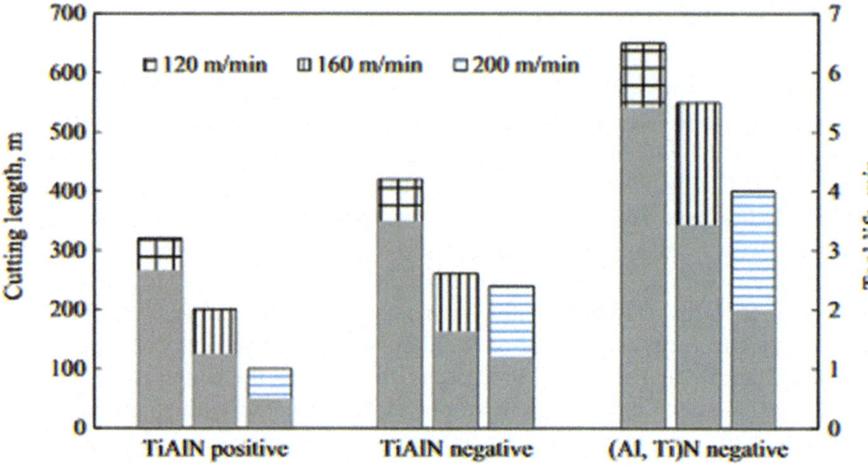

Fig. 2.18 Tool life comparison between TiAlN positive, TiAlN negative, and (Al, Ti)N negative at various cutting speeds [67]

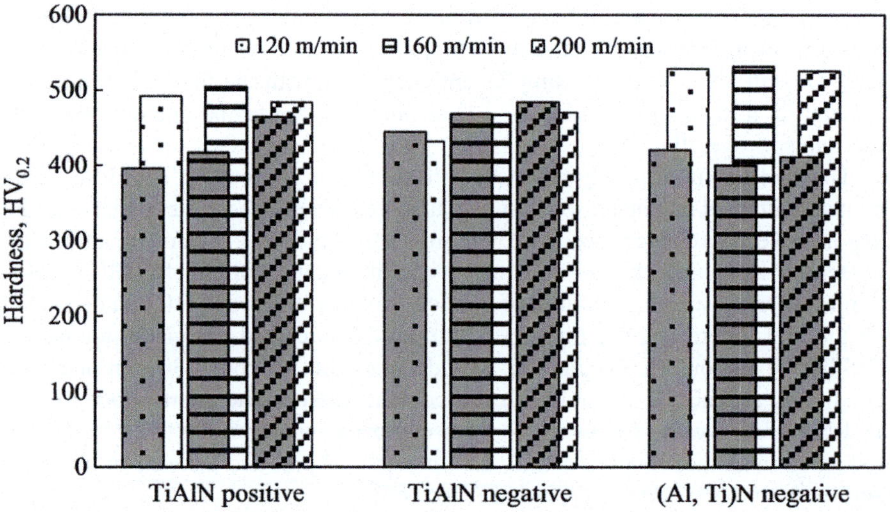

Fig. 2.19 Hardness improvement comparison between TiAlN positive, TiAlN negative, and (Al, Ti)N negative at various cutting speeds [67]

2.4.3 *Grinding of Inconel*

As in the other HSM processes, CBN and PCBN tools can be applied in high-speed grinding due to their benefits in machining at higher speeds. When comparing machining using these tools inside the conventional range and high-speed grinding range, the power consumption during the process is an important factor. High-speed grinding leads to higher power consumption at around 55–65% loss at a grinding wheel speed of 140 m/s, whereas the loss for conventional machining is around 20–30%. Despite high power consumption, CBN tools have significant advantages in high-speed grinding: The lowest grinding force and specific grinding energy are achieved when the machining is conducted at 120 m/s, which also ensures that the temperature is below 100 °C; in addition, the surface roughness reduces to its minimum, below 0.4 μm at grinding speeds of 100–120 m/s [68].

Self-sharpening is an important quality of PCBN, which has been studied with regard to high-speed grinding in the last several years. During machining with PCBN, fractures occur due to crack propagation caused by non-uniform deformation and due to microfractures occurring in the area of the cutting edge caused by stress concentration. Microfractures can be beneficial because they create a large number of micro-cutting edges that divide the working zone into multiple layers leading to a variation in chip propagation and therefore self-sharpen the tool. However, the number of cutting edges should be balanced because too few lead to too large size of chips, while too many lead to higher energy losses. The balance between these two characteristics is achieved when around 40–50% of the regions are affected by microfractures. In contrast to that, CBN forms macrofractures instead, which lead to smooth slipping surfaces. Therefore, CBN tools result in a smaller average surface roughness of around 0.20 μm, whereas PCBN results in it being around 0.30 μm on average. Nevertheless, since PCBN has a smaller grinding force ratio compared with CBN, it means that grain sharpness in PCBN is higher [69, 70].

Ultra-high-speed grinding could lead to even faster machining times and higher productivity. Therefore, it is necessary to understand the mechanisms behind material removal and chip formation at such high speeds to better foresee the potential effects, especially in difficult-to-cut materials such as Inconel 718. Qiu et al. [71] have conducted a study combining the experimental results with the results from FEA to observe the effects of cutting speeds on chip formation at grinding wheel speeds up to 1000 m/s. As the grinding speed grows from rest, the chips first appear continuous; at 80 m/s, with the increase in grinding force fluctuation caused by shear slips in the material, the chips become continuously serrated; when the speed increases past 240 m/s, the degree of serration increases with the reduction of space between each tooth; lastly, when the speed increases beyond 400 m/s, the chip morphology changes from serrated to banded continuity, which leads to reduction in the grinding force and improved machinability. Figure 2.20 shows the evolution of chip morphology.

2.4.4 Drilling of Inconel

High-speed drilling conditions are difficult to achieve when machining Inconel 718 due to high heat generation and rapid tool wear not allowing to utilize the drilling tools at high cutting speeds. Neo et al. [72] proposed a new PCBN-tipped gun drill to combat these challenges. The PCBN layer was installed on the cutting tip and bearing pads. The cutting speed up to 75.4 m/min was tested. The modified

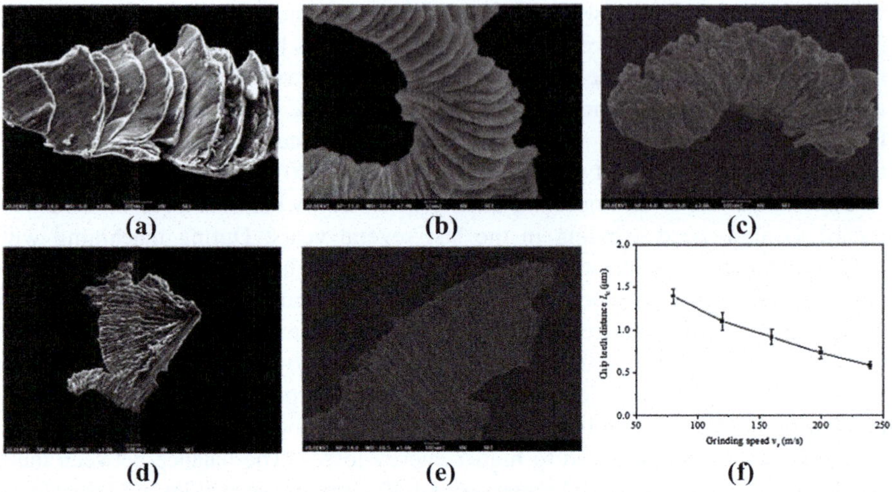

Fig. 2.20 Evolution of serrated chip appearance with an increasing grinding wheel speed: (**a**) 80 m/s; (**b**) 120 m/s; (**c**) 160 m/s; (**d**) 200 m/s; (**e**) 240 m/s; (**f**) changes in shear slip distance [71]

PCBN drill showed a significantly improved performance at high speeds, reduced drilling forces and torques, and smaller tool wear. At higher cutting speeds, it also reduces surface roughness by around four to five times depending on hole depth (e.g., from 1.25 to 0.25 mm for a 160–200 mm deep hole) and improves hole straightness by up to two times (e.g., ~0.075 versus 0.15 deviation/mm at 200-mm hole depth) compared with tungsten carbide drill, which is not able to achieve high cutting speeds.

2.5 Conclusions and Outlook

In recent years, there have been significant developments in HSM of aerospace alloys such as AA7050, AA7075, Ti-6Al-4V, and Inconel 718. However, as better mechanical and thermal properties are sought after in the aerospace industry at the same time as the need for productivity and efficiency grows, the machining challenges still persist: preventing tool wear, managing high machining temperatures, improving surface integrity, etc. In order to combat these challenges while keeping up with the demand of the industry, continuous research and improvement of HSM are carried out. In HSM of aluminum alloys, which do not pose as many challenges in the machining process compared with the other alloys presented, the research focuses mainly on the improvement of efficiency and productivity of the machining process with investigations into optimal machining parameters, better cooling, and additional processes such as shot peening improving the resulting qualities of the alloy. In difficult-to-cut alloys such as Ti-6Al-4V and Inconel 718, the research primarily focuses on reducing tool wear via the employment of new tool materials, coating methods, or cooling conditions. In addition, new machining techniques such as vibration assistance and integration into additive manufacturing have also been investigated to achieve better properties of resulting products.

As sustainability becomes an important factor in the manufacturing industry, HSM methods under dry conditions and MQL are being developed to reduce the waste of the coolant fluid. Future research should continue to focus on enhancing the sustainability aspects of the HSM processes with a focus on aerospace materials. One important aspect of HSM is formation of residual stresses at the subsurface of the machined aerospace materials. Residual stress can significantly influence the surface integrity and mechanical properties and thus product performance of the final part [73]. Therefore, future researchers should focus on investigating and mitigating residual stress formation during HSM of aerospace materials. Overall, the focus of HSM in aerospace alloy machining is expected to continue to evolve with an emphasis on efficiency, productivity, predictive models, material properties innovation, and sustainability. Future research on HSM should consider machining of other aerospace materials including CFRP composites, as well as various metal matrix and ceramic matrix composites. With the growth of additive manufacturing of CFRP and other aerospace alloys, HSM of three-dimensional (3D)-printed CFRP, Ti-6Al-4V, and Inconel alloys is already attracting attention of researchers [74–76]. In addition to HSM of these additively manufactured aerospace materials, sustainability assessments have also been considered in recent studies [76]. Finally, the application of artificial intelligence and machine learning in the advanced machining processes has grown significantly in recent years, although most of these studies focused on additive manufacturing [77]. Therefore, there is a huge opportunity of growth on the application of artificial intelligence and machine learning in HSM of aerospace materials, which could significantly improve the machining performance of various HSM processes and reduce the cost of machining these difficult-to-cut materials, by artificial intelligence and machine learning-based optimization techniques.

References

1. Boyer RR, Cotton JD, Mohaghegh M, Schafrik RE. Materials considerations for aerospace applications. MRS Bull. 2015;40:1055–66. https://doi.org/10.1557/mrs.2015.278.
2. Wang D. Research on surface integrity and its influencing factors in the high-speed cutting of typical aluminum/titanium/nickel alloys: a review. Int J Adv Manuf Technol. 2023;127:4915–42. https://doi.org/10.1007/s00170-023-11808-0.
3. Schulz H, Moriwaki T. High-speed machining. CIRP Ann. 1992;41:637–43. https://doi.org/10.1016/S0007-8506(07)63250-8.
4. Hon K, Baharudin BHT. The impact of high speed machining on computing and automation. Int J Automat Comput. 2006;3:63–8. https://doi.org/10.1007/s11633-006-0063-3.
5. Smith S, Tlusty J. Current trends in high-speed machining. J Manuf Sci Eng. 1997;119:664–6. https://doi.org/10.1115/1.2836806.

6. Li B, Ni J, Yang J, Liang SY. Study on high-speed grinding mechanisms for quality and process efficiency. Int J Adv Manuf Technol. 2014;70:813–9. https://doi.org/10.1007/s00170-013-5297-y.
7. Jain A, Bajpai V. Chapter 1—introduction to high-speed machining (HSM). In: Gupta K, Paulo Davim J, editors. High speed machining. Academic Press; 2020. p. 1–25.
8. Salguero J, Batista M, del Sol I, Vazquez JM. Chapter 5—high-speed machining of aluminum alloys for aircraft structures. In: Gürgen S, Pruncu CI, editors. Modern manufacturing processes for aircraft materials. Elsevier; 2024. p. 91–123.
9. Hourmand M, Sarhan AAD, Sayuti M, Hamdi M. A comprehensive review on machining of titanium alloys. Arab J Sci Eng. 2021;46:7087–123. https://doi.org/10.1007/s13369-021-05420-1.
10. Warsi SS, Zahid T, Elahi H, Liaqait RA, Bibi S, Gillani F, Ghafoor U. Sustainability-based analysis of conventional to high-speed machining of Al 6061-T6 alloy. Appl Sci. 2021;11:9032. https://doi.org/10.3390/app11199032.
11. Bałon P, Rejman E, Smusz R, Szostak J, Kiełbasa B. Implementation of high speed machining in thin-walled aircraft integral elements. Open Eng. 2018;8:162–9. https://doi.org/10.1515/eng-2018-0021.
12. Bäker M. Finite element simulation of high-speed cutting forces. J Mater Process Technol. 2006;176:117–26. https://doi.org/10.1016/j.jmatprotec.2006.02.019.
13. Perez I, Madariaga A, Cuesta M, Garay A, Arrazola PJ, Ruiz JJ, Rubio FJ, Sanchez R. Effect of cutting speed on the surface integrity of face milled 7050-T7451 aluminium workpieces. Procedia CIRP. 2018;71:460–5. https://doi.org/10.1016/j.procir.2018.05.034.
14. Huang X, Xu J, Chen M, Ren F. Finite element modeling of high-speed milling 7050-T7451 alloys. Procedia Manuf. 2020;43:471–8. https://doi.org/10.1016/j.promfg.2020.02.186.
15. Ping Z, Yue X, Shuangfeng H, Ailing S, Baoshun L, Xiao Y. Experiment and simulation on the high-speed milling mechanism of aluminum alloy 7050-T7451. Vacuum. 2020;182:109778. https://doi.org/10.1016/j.vacuum.2020.109778.
16. Zhong Z, Ai X, Liu Z, Liu J, Xu Q. Surface morphology and microcrack formation for 7050-T7451 aluminum alloy in high speed milling. Int J Adv Manuf Technol. 2015;78:281–96. https://doi.org/10.1007/s00170-014-6605-x.
17. Zhang X, Ma J, Jia Z, Song D. Machining parameter optimisation for aviation aluminium-alloy thin-walled parts in high-speed milling. Int J Mach Mach Mater. 2018;20:180–92. https://doi.org/10.1504/IJMMM.2018.090546.
18. Lukic D, Cep R, Vukman J, Antic A, Djurdjev M, Milosevic M. Multi-criteria selection of the optimal parameters for high-speed machining of aluminum alloy Al7075 thin-walled parts. Metals. 2020;10:1570. https://doi.org/10.3390/met10121570.
19. Wang Z, Cao Y, Yao H, Kou F. Dynamic simulation and experimental study of cutting force by rake angle of multi-axis high-speed ball-end milling tool. Int J Adv Manuf Technol. 2022;122: 377–90. https://doi.org/10.1007/s00170-022-09528-y.
20. Żyłka Ł, Flejszar R, Lajmert P. Influence of cutting-edge microgeometry on cutting forces in high-speed milling of 7075 aluminum alloy. Materials. 2023;16:3859. https://doi.org/10.3390/ma16103859.
21. Xiuli F, Wenxing L, Yongzhi P, Wentao L. Morphology evolution and micro-mechanism of chip formation during high-speed machining. Int J Adv Manuf Technol. 2018;98:165–75. https://doi.org/10.1007/s00170-017-0411-1.
22. Wang B, Liu Z, Song Q, Wan Y, Ren X. A modified Johnson–cook constitutive model and its application to high speed machining of 7050-T7451 aluminum alloy. J Manuf Sci Eng. 2018;141 https://doi.org/10.1115/1.4041915.
23. Wang Z, Cao Y, Gorbachev S, Kuzin V, He W, Guo J. Research on conventional and high-speed machining cutting force of 7075-T6 aluminum alloy based on finite element modeling and simulation. Metals. 2022;12:1395. https://doi.org/10.3390/met12081395.

24. Paturi UMR, Narala SKR. Constitutive flow stress formulation for aeronautic aluminum alloy AA7075-T6 at elevated temperature and model validation using finite element simulation. Proc Inst Mech Eng Part L: J Mater Des Appl. 2016;230:994–1004. https://doi.org/10.1177/1464420715591860.
25. Damous M, Zeroudi N, Chellil A, Ikkache K. Prediction of a chatter stability of thin-walled parts during high-speed milling considering a variation of dynamic characteristics based on surface roughness measurement. Int J Adv Manuf Technol. 2023;127:5729–44. https://doi.org/10.1007/s00170-023-11945-6.
26. Zhang P, Yue X, Gao Y, Wang S, Sun Y, Zhou H, Zhang J. Research on the surface corrosion behavior of 7075-T6 aluminum alloy during high-speed machining and particle inclusion water jet composite reinforcement. Vacuum. 2024;219:112700. https://doi.org/10.1016/j.vacuum.2023.112700.
27. Lu W, Zong C, Ni C, Yu X, Liu D. Study on the surface integrity of 7050 aluminum alloy with different crystal orientations during high-speed machining. Int J Adv Manuf Technol. 2023;125:661–78. https://doi.org/10.1007/s00170-022-10717-y.
28. Imbrogno S, Rinaldi S, Suarez AG, Arrazola PJ, Umbrello D. High speed machinability of the aerospace alloy AA7075 T6 under different cooling conditions. AIP Conf Proc. 2018;1960: 070013. https://doi.org/10.1063/1.5034909.
29. Imbrogno S, Rotella G, Rinaldi S. Surface and subsurface modifications of AA7075-T6 induced by dry and cryogenic high speed machining. Int J Adv Manuf Technol. 2020;107:905–18. https://doi.org/10.1007/s00170-020-05108-0.
30. Okafor AC. Chapter 5—cooling and machining strategies for high speed milling of titanium and nickel super alloys. In: Gupta K, Paulo Davim J, editors. High speed machining. Academic Press; 2020. p. 127–61.
31. Matuszak J, Zaleski K, Zyśko A. Investigation of the impact of high-speed machining in the milling process of titanium alloy on tool Wear, surface layer properties, and fatigue life of the machined object. Materials. 2023;16:5361. https://doi.org/10.3390/ma16155361.
32. Ullah I, Zhang S, Zhang Q, Li B. Microstructural and mechanical property investigation of machined surface layer in high-speed milling of Ti-6Al-4V alloy. Int J Adv Manuf Technol. 2021;116:1707–19. https://doi.org/10.1007/s00170-021-07568-4.
33. Liu H, Zhang J, Xu X, Zhao W. Experimental study on fracture mechanism transformation in chip segmentation of Ti-6Al-4V alloys during high-speed machining. J Mater Process Technol. 2018;257:132–40. https://doi.org/10.1016/j.jmatprotec.2018.02.040.
34. Ullah I, Zhang S, Zhang Q, Wang R. Numerical investigation on serrated chip formation during high-speed milling of Ti-6Al-4V alloy. J Manuf Process. 2021;71:589–603. https://doi.org/10.1016/j.jmapro.2021.09.056.
35. Kuram E. The effect of monolayer TiCN-, AlTiN-, TiAlN-and two layers TiCN + TiN- and AlTiN + TiN-coated cutting tools on tool wear, cutting force, surface roughness and chip morphology during high-speed milling of Ti6Al4V titanium alloy. Proc Inst Mech Eng B J Eng Manuf. 2018;232:1273–86. https://doi.org/10.1177/0954405416666905.
36. Wang F, Ji K, Guo Z. Microstructural analysis of failure progression for coated carbide tools during high-speed milling of Ti-6Al-4V. Wear. 2020;456–457:203356. https://doi.org/10.1016/j.wear.2020.203356.
37. Khatri A, Jahan MP, Ma J. Assessment of tool wear and microstructural alteration of the cutting tools in conventional and sustainable slot milling of Ti-6Al-4V alloy. Int J Adv Manuf Technol. 2019;105:2799–814. https://doi.org/10.1007/s00170-019-04520-5.
38. Khatri A, Jahan MP. Investigating the microstructure and morphology of chips in dry, flood coolant and MQL machining of Ti-6Al-4V alloy. Int J Sustain Manuf. 2020;4:234–50. https://doi.org/10.1504/IJSM.2020.107136.
39. Khosravi J, Azarhoushang B, Barmouz M, Bösinger R, Zahedi A. High-speed milling of Ti6Al4V under a supercritical CO2 + MQL hybrid cooling system. J Manuf Process. 2022;82:1–14. https://doi.org/10.1016/j.jmapro.2022.07.061.

40. Han G, Ye Z, Xu J, Ma Y, Xu C, Zhao X, Yu L, Feng P, Feng F. Investigation on cutting forces and tool wear in high-speed milling of Ti-6Al-4V assisted by longitudinal torsional ultrasonic vibrations. Int J Adv Manuf Technol. 2023;129:783–99. https://doi.org/10.1007/s00170-023-12306-z.
41. Zhang J, Ling L, Wang Q-Y, Huang X-F, Kang X-Z, Tao G-B, Cao H-J. Surface quality investigation in high-speed dry milling of Ti-6Al-4V by using 2D ultrasonic-vibration-assisted milling platform. Adv Manuf. 2024; https://doi.org/10.1007/s40436-023-00473-x.
42. Zhang M, Zhang D, Guo H, Gao Z, Geng D, Liu J, Jiang X. High-speed rotary ultrasonic elliptical milling of Ti-6Al-4V using high-pressure coolant. Metals. 2020;10:500. https://doi.org/10.3390/met10040500.
43. Tao Z, An Q, Liu G, Chen M. A novel method for tool condition monitoring based on long short-term memory and hidden Markov model hybrid framework in high-speed milling Ti-6Al-4V. Int J Adv Manuf Technol. 2019;105:3165–82. https://doi.org/10.1007/s00170-019-04464-w.
44. Zhang H, Dang J, Ming W, Xu X, Chen M, An Q. Cutting responses of additive manufactured Ti6Al4V with solid ceramic tool under dry high-speed milling processes. Ceram Int. 2020;46: 14536–47. https://doi.org/10.1016/j.ceramint.2020.02.253.
45. D'Mello G, Srinivasa Pai P, Prashanth A. Surface roughness analysis in high speed turning of Ti-6Al-4V using coated carbide inserts: experimental and modeling studies. Tribol Ind. 2018;40:457–76. https://doi.org/10.24874/ti.2018.40.03.12.
46. Liang X, Liu Z, Wang B, Hou X. Modeling of plastic deformation induced by thermo-mechanical stresses considering tool flank wear in high-speed machining Ti-6Al-4V. Int J Mech Sci. 2018;140:1–12. https://doi.org/10.1016/j.ijmecsci.2018.02.031.
47. da Silva LR, da Silva OS, dos Santos FV, Duarte FJ, Veloso GV. Wear mechanisms of cutting tools in high-speed turning of Ti6Al4V alloy. Int J Adv Manuf Technol. 2019;103:37–48. https://doi.org/10.1007/s00170-019-03519-2.
48. Chowdhury MSI, Bose B, Yamamoto K, Shuster LS, Paiva J, Fox-Rabinovich GS, Veldhuis SC. Wear performance investigation of PVD coated and uncoated carbide tools during high-speed machining of TiAl6V4 aerospace alloy. Wear. 2020;446–447:203168. https://doi.org/10.1016/j.wear.2019.203168.
49. Chang K, Zheng G, Cheng X, Xu R, Li Y, Yu Z, Yang X. Surface integrity evolution and wear evolution of the micro-blasted coated tool in high-speed turning of Ti6Al4V. Int J Adv Manuf Technol. 2021;115:603–16. https://doi.org/10.1007/s00170-021-07227-8.
50. Yi J, Zhou W, Deng Z. Experimental study and numerical simulation of the intermittent feed high-speed grinding of TC4 titanium alloy. Metals. 2019;9:802. https://doi.org/10.3390/met9070802.
51. Naskar A, Choudhary A, Paul S. Surface generation in ultrasonic-assisted high-speed superabrasive grinding under minimum quantity cooling lubrication with various fluids. Tribol Int. 2021;156:106815. https://doi.org/10.1016/j.triboint.2020.106815.
52. Prasad BS, Sai Ravi Kiran DS. Experimental investigation to optimize tool performance in high-speed drilling: a comparative study. J Braz Soc Mech Sci Eng. 2019;41:535. https://doi.org/10.1007/s40430-019-2049-4.
53. Hari Chealvan S, Krishnaraj V, Rahim EA. High speed cryogenic drilling of Ti-6Al-4V alloy under high pressure coolant conditions. Proc Inst Mech Eng B J Eng Manuf. 2022;236:1633–42. https://doi.org/10.1177/09544054221080025.
54. Guo D, Guo X, Zhang K, Chen Y, Zhou C, Gai L. Improving cutting performance of carbide twist drill combined internal cooling and micro-groove textures in high-speed drilling Ti6Al4V. Int J Adv Manuf Technol. 2019;100:381–9. https://doi.org/10.1007/s00170-018-2733-z.
55. Ulutan D, Ozel T. Machining induced surface integrity in titanium and nickel alloys: a review. Int J Mach Tools Manuf. 2011;51:250–80. https://doi.org/10.1016/j.ijmachtools.2010.11.003.
56. De Bartolomeis A, Newman ST, Jawahir IS, Biermann D, Shokrani A. Future research directions in the machining of Inconel 718. J Mater Process Technol. 2021;297:117260. https://doi.org/10.1016/j.jmatprotec.2021.117260.

57. Ma J, Jia Z, He G, Liu Z, Zhao X, Qin F. Influence of cutting tool geometrical parameters on tool wear in high-speed milling of Inconel 718 curved surface. Proc Inst Mech Eng B J Eng Manuf. 2019;233:18–30. https://doi.org/10.1177/0954405417716495.
58. Gueli M, Ma J, Cococcetta N, Pearl D, Jahan MP. Experimental investigation into tool wear, cutting forces, and resulting surface finish during dry and flood coolant slot milling of Inconel 718. Procedia Manuf. 2021;53:236–45. https://doi.org/10.1016/j.promfg.2021.06.026.
59. Zhang G, Zhang J, Fan G, Xu C, Du J. The effect of chip formation on the cutting force and tool wear in high-speed milling Inconel 718. Int J Adv Manuf Technol. 2023;127:335–48. https://doi.org/10.1007/s00170-023-11551-6.
60. Rinaldi S, Umbrello D, Rotella G, Prete AD. A physically based model to predict microstructural modifications in *Inconel 718* high speed machining. Procedia Manuf. 2020;47:487–92. https://doi.org/10.1016/j.promfg.2020.04.344.
61. Grguraš D, Kern M, Pušavec F. Suitability of the full body ceramic end milling tools for high speed machining of nickel based alloy Inconel 718. Procedia CIRP. 2018;77:630–3. https://doi.org/10.1016/j.procir.2018.08.190.
62. Halim NHA, Haron CHC, Ghani JA, Azhar MF. Tool wear and chip morphology in high-speed milling of hardened Inconel 718 under dry and cryogenic CO2 conditions. Wear. 2019;426–427:1683–90. https://doi.org/10.1016/j.wear.2019.01.095.
63. Sommer D, Hornung S, Esen C, Hellmann R. Surface roughness optimization of hybrid PBF-LB/M-built Inconel 718 using in situ high-speed milling. Int J Adv Manuf Technol. 2024; https://doi.org/10.1007/s00170-024-13382-5.
64. Tu L, Lin L, Liu C, Zheng T, Deng Y, Han L, An Q, Ming W, Chen M. Tool wear characteristics analysis of cBN cutting tools in high-speed turning of Inconel 718. Ceram Int. 2023;49:635–58. https://doi.org/10.1016/j.ceramint.2022.09.034.
65. Díaz-Álvarez J, Criado V, Miguélez H, Cantero JL. PCBN performance in high speed finishing turning of Inconel 718. Metals. 2018;8:582. https://doi.org/10.3390/met8080582.
66. Cantero JL, Díaz-Álvarez J, Infante-García D, Rodríguez M, Criado V. High speed finish turning of Inconel 718 using PCBN tools under dry conditions. Metals. 2018;8:192. https://doi.org/10.3390/met8030192.
67. Zhang B, Njora MJ, Sato Y. High-speed turning of Inconel 718 by using TiAlN- and (Al, Ti) N-coated carbide tools. Int J Adv Manuf Technol. 2018;96:2141–7. https://doi.org/10.1007/s00170-018-1765-8.
68. Dai C-W, Ding W-F, Zhu Y-J, Xu J-H, Yu H-W. Grinding temperature and power consumption in high speed grinding of Inconel 718 nickel-based superalloy with a vitrified CBN wheel. Precis Eng. 2018;52:192–200. https://doi.org/10.1016/j.precisioneng.2017.12.005.
69. Rao Z, Ding W, Zhu Y, Su H. Understanding the self-sharpening characteristics of polycrystalline cubic boron nitride super-abrasive in high-speed grinding of Inconel 718. Ceram Int. 2019;45:13324–33. https://doi.org/10.1016/j.ceramint.2019.04.024.
70. Wang J, Ding W, Zhu Y, Fu Y. Micro-fracture variation and grinding performance of PCBN superabrasive grains in high-speed grinding. Int J Mech Sci. 2019;160:15–25. https://doi.org/10.1016/j.ijmecsci.2019.06.021.
71. Qiu B, Zhu Y, Ding W. An investigation on material removal mechanism in ultra-high-speed grinding of nickel-based superalloy: three-dimensional simulation and experimental verification. Int J Adv Manuf Technol. 2020;110:919–33. https://doi.org/10.1007/s00170-020-05922-6.
72. Neo DWK, Liu K, Kumar AS. High throughput deep-hole drilling of Inconel 718 using PCBN gun drill. J Manuf Process. 2020;57:302–11. https://doi.org/10.1016/j.jmapro.2020.06.043.
73. Bastola N, Jahan MP, Rangasamy N, Rakurty CS. A review of the residual stress generation in metal additive manufacturing: analysis of cause. Meas Eff Prev Micro. 2023;14(7):1480. https://doi.org/10.3390/mi14071480.
74. Cococcetta N, Jahan MP, Schoop J, Ma J, Pearl D, Hassan M. Post-processing of 3D printed thermoplastic CFRP composites using cryogenic machining. J Manuf Process. 2021;68:332–46. https://doi.org/10.1016/j.jmapro.2021.05.054.

75. Cococcetta NM, Pearl D, Jahan MP, Ma J. Investigating surface finish, burr formation, and tool wear during machining of 3D printed carbon fiber reinforced polymer composite. J Manuf Process. 2020;56:1304–16. https://doi.org/10.1016/j.jmapro.2020.04.025.
76. Alimuzzaman SM, Jahan MP, Rakurty CS, Rangasamy N, Ma J. Cutting fluids in metal AM: a review of sustainability and efficiency. J Manuf Process. 2023;106:51–87. https://doi.org/10.1016/j.jmapro.2023.09.075.
77. Rahman MA, Saleh T, Jahan MP, McGarry C, Chaudhari A, Huang R, Tauhiduzzaman M, Ahmed A, Mahmud AA, Bhuiyan MS, Khan MF, Alam MS, Shakur MS. Review of intelligence for additive and subtractive manufacturing: current status and future prospects. Micromachines. 2023;14(3):508. https://doi.org/10.3390/mi14030508.

Chapter 3
Ultrasonic Vibration-Assisted Machining with Minimum Quantity Lubrication for Aerospace Materials

Erkin Duman, Yusuf Furkan Yapan, Alper Uysal, and Mehmet Alper Sofuoğlu

3.1 Introduction

The aerospace and aviation industry has experienced a significant surge in component demand recently, driven by the accelerated production of aircraft, space shuttles, and satellites. Boeing's "Current Market Outlook 2016–2035" report predicts a doubling of the commercial jet fleet by 2036, anticipating a market value of approximately US$6.1 trillion. In addition, research by Madhwal et al. [1] indicates a projected 60% increase in aircraft fleets in Asia, with Europe and America likely to see considerable fleet replacements in the next 5 years. Consequently, industries are actively exploring efficient manufacturing strategies for aerospace components to meet this escalating demand and maximize profitability. Manufacturers in the aerospace sector are rigorously working to optimize aircraft weight and design for enhanced fuel efficiency. This effort has spurred ongoing research and development to introduce materials that possess exceptional thermochemical properties, such as being lightweight, strong, having low thermal expansion coefficients, and offering corrosion resistance. Recently, the aerospace and aviation sectors have seen the introduction of innovative metallic alloys and composites. These developments are geared toward overcoming challenges in reducing aircraft weight relative to fuel consumption and enhancing aerodynamic efficiency. In their study, Saoubi et al. [2]

E. Duman
Department of Machine and Metal Technology, Gedik Vocational School, Istanbul Gedik University, Istanbul, Turkey

Y. F. Yapan · A. Uysal
Department of Mechanical Engineering, Yıldız Technical University, Istanbul, Turkey

M. A. Sofuoğlu (✉)
Department of Mechanical Engineering, Eskişehir Osmangazi University, Eskişehir, Turkey
e-mail: asofuoglu@ogu.edu.tr

S. Gürgen (ed.), *Advances in Manufacturing for Aerospace Alloys*, Sustainable Aviation, https://doi.org/10.1007/978-3-031-64455-9_3

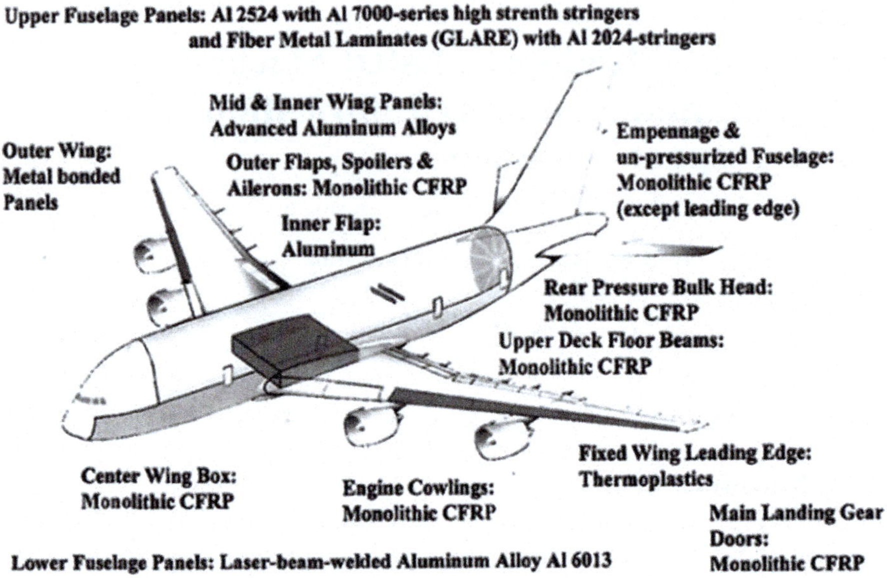

Fig. 3.1 Materials used in Airbus A380 [3]

explored the impact of these advanced materials in the construction of contemporary aerostructures. They observed that composites constitute about 50% of the material weight in aircraft such as the Bombardier CS300, Boeing B787, and Airbus A350XWB. The authors anticipate a growing demand for advanced materials, including superalloys and composites, in the aerospace manufacturing industry going forward. Fig. 3.1 illustrates a schematic showing the materials used in Airbus A380.

Heat-resistant materials with high fusion points are essential in crafting components for aeroengines. These specialized alloys are divided into four primary categories: nickel, cobalt, iron-based (like high chromium stainless steel), and titanium alloys. Around two-thirds of all such alloy production are utilized by the aerospace sector, predominantly for jet engine parts and associated elements, particularly in areas of aircraft engines and land-based turbines that encounter high temperatures. Superalloys are preferred for their ability to retain strong mechanical and chemical characteristics under high heat, making them suitable for both moving and static parts in the heated segments of jet engines. Using these alloys in components results in lighter, smaller parts compared with the traditional steel, thus significantly reducing fuel consumption and decreasing environmental pollution. For aeroengines, each kilogram of weight saved can mean fuel cost reductions of about US$150,000 during the engine's operational life. The other third of superalloy production serves the chemical, medical, and structural fields, where high-temperature endurance and remarkable corrosion resistance are required. These varied uses emphasize the vital role and flexibility of superalloys in multiple industries [4].

Titanium alloys have been developed to meet the demand for a class of materials that are both strong and lightweight, particularly for the manufacture of aircraft engines and airframes. This is largely due to their exceptional strength-to-density ratios. In addition to their remarkable strength, titanium alloys also offer excellent corrosion resistance, which translates into cost savings as there is less need for protective coatings, such as paint, that would typically be required for steel. In aircraft frame construction, these alloys prove to be especially beneficial where temperatures surpass 130 °C, exceeding the usual limit for aluminum alloys. Within aeroengines, the use of titanium alloys is widespread in both low-pressure and high-pressure compressors. They play a crucial role in producing parts that endure significant centrifugal forces, like disks and blades, particularly those with smaller flow diameters, and in parts that function under extreme fatigue conditions. Beyond aerospace, titanium alloys have found a widespread use in various other industries. This is attributed to their good corrosion resistance and the ability to develop a range of mechanical and physical properties tailored to specific applications [5–7].

Aluminum alloys are a cornerstone in the aerospace industry, owing to their excellent strength-to-weight ratio, making them ideal for aircraft design. One of the most commonly used alloys in this sector is the AA2024, known for its high strength and significant resistance to fatigue, making it perfect for aircraft structural components [8]. The AA7075, another popular choice, stands out for its superior strength, which is beneficial for highly stressed structural parts and is often used in manufacturing aircraft wings and fuselages [9]. The AA6061, recognized for its versatility, weldability, and corrosion resistance, finds applications in aircraft fuselage and wings due to its good balance of strength and weight [10]. These alloys not only contribute to the structural integrity and efficiency of aircraft but also play a crucial role in advancements in aerospace engineering, focusing on creating lighter and more fuel-efficient aircraft. This ongoing evolution in the use of aluminum alloys is pivotal in meeting the aerospace industry's demands for sustainability and performance. Table 3.1 lists the mechanical properties of some aluminum alloys.

Magnesium alloys, renowned for their low density and high strength-to-weight ratio, are increasingly used in the aerospace industry to enhance fuel efficiency and reduce aircraft weight. For instance, the magnesium alloy AZ31B is utilized for its good strength and ductility, often seen in aircraft seat frames and interior components [12]. Another alloy, AZ91D, is known for its excellent castability and good

Table 3.1 Mechanical properties of some aluminum alloys [11]

Al alloys	Yield strength (MPa)	Ultimate tensile strength (MPa)	Elongation at break (%)
AA2025-T6	255	400	19
AA2090-T83	520	550	6
AA7075	503	572	11
AA7475-T651	510	586	13
AA8090-T852	410	480	7

mechanical properties, making it suitable for electronic equipment cases in aircraft [13]. High-temperature magnesium alloys such as WE43 and Elektron 21 (EV31A) are employed in gearbox housings and other engine components due to their outstanding creep resistance and high-temperature performance [14]. In addition, ZK60, with its high strength and good formability, finds application in aerospace components where higher strength is required [15]. These magnesium alloys contribute significantly to weight reduction in aircraft, thereby improving fuel efficiency and reducing environmental impact. However, their application requires careful consideration of factors like flammability and corrosion resistance. The use of protective coatings and advanced alloying techniques helps mitigate these issues, allowing for the safe and effective use of magnesium alloys in the aerospace sector.

Steel plays a critical role in the aerospace industry, particularly where high strength, durability, and resistance to environmental stress are required. One prominent example is the use of maraging steels, known for their ultra-high strength and toughness, in aerospace components such as landing gear, helicopter rotor shafts, and rocket motor casings [16]. These steels, like the 18Ni (250) grade, achieve their properties through age-hardening, providing an excellent combination of strength and malleability [17]. In addition, stainless steels, such as the 300 series (especially 304 and 316 types), are widely used for their corrosion resistance and strength at high temperatures, finding applications in engine components, exhaust systems, and other critical parts exposed to harsh conditions [18]. High-speed tool steels, like M2, are utilized for their wear resistance and hardness, essential for components like turbine blades [19]. The aerospace industry also employs alloy steels such as 4130 and 4340 for their balance of toughness, strength, and fatigue resistance in structural components [20]. These steels are integral to the aerospace industry, contributing to the safety, efficiency, and longevity of aircraft and space vehicles.

Approximately 50% of the alloys used in aeroengine construction by weight are nickel-based. These alloys are preferred over denser materials like steel due to their superior strength-to-weight ratio. Nickel-based alloys are especially valuable in demanding aerospace environments because of their capacity to maintain high levels of resistance to various stresses. They are particularly adept at withstanding mechanical and thermal fatigue, creep, erosion, corrosion, and mechanical and thermal shock, even at elevated temperatures. This combination of properties makes them ideal for critical aerospace applications where durability and reliability under extreme conditions are paramount [21].

Iron-based superalloys, though not as robust at high temperatures compared with nickel alloys, find their primary application in chemical processing. These alloys, despite their relatively lower strength at elevated temperatures, are suitable for environments where chemical resistance is crucial. On the other hand, cobalt-based alloys, known for their higher melting points than both nickel- and iron-based counterparts, are extensively utilized in applications like turbine vanes. Their preference in such applications is also due to their exceptional corrosion resistance, making them a vital material in environments where both high temperature and corrosion resistance are essential [22].

Composite materials have transformed the aerospace industry by providing a remarkable blend of high strength-to-weight ratios, durability, and corrosion resistance, playing a pivotal role in modern aircraft design. These materials consist of two or more distinct constituents, such as carbon fibers embedded in a polymer matrix or ceramic particles in a metal matrix, offering enhanced structural performance while reducing overall weight. They have become indispensable in various aerospace applications, including aircraft fuselages, wings, and engine components, contributing to increased fuel efficiency, extended service life, and improved aircraft performance [23]. Composites are a preferred choice for aerospace engineers due to their capacity to meet rigorous requirements for strength and durability while minimizing weight, thus advancing aerospace technology [24].

Titanium- and nickel-based superalloys, recognized for their exceptional high-temperature mechanical and chemical attributes, are extensively utilized in the aerospace, marine, and chemical sectors. However, these alloys' superior thermal properties often result in challenges during machining, including poor surface quality. Factors such as fluctuating chip thickness, intense thermal stress, elevated pressure, material rebound, and residual stress contribute to increased tool degradation and suboptimal surface finish. In contrast, the incorporation of composite materials, notably carbon fiber reinforced polymers (CFRPs), in aircraft production enhances performance through lightness and strength, simultaneously reducing life cycle costs, maintenance expenses, and production time while ensuring quality and cost-efficiency. Nonetheless, machining difficulties of these composites can impair their durability and fatigue resistance. Particularly in drilling, delamination on hole surfaces is a notable issue. To mitigate these issues, advanced cooling methods and hybrid machining techniques are being employed [25, 26].

In the field of aerospace manufacturing, ensuring the reliability and safety of components is paramount. This is particularly true in the context of upgrading aerospace engines, where the main structures typically incorporate robust high-strength alloys. The precision in machining process directly impacts the component usability. A crucial aspect in this regard is the assessment of machined surface quality, focusing primarily on surface integrity. Consequently, enhancing the surface integrity of aerospace parts has emerged as a significant area of interest in the industry [26–29].

Turning operations comprise cylindrical turning, face turning, knurling, drilling, grooving, threading, reaming, contouring, taper turning, and special applications. According to the literature, it is a well-established fact that the ultrasonic vibration-assisted (UVA) machining process utilizing the minimum quantity lubrication (MQL) methods (UVA & MQL) of aerospace alloys includes 43% turning (UVAT & MQL), 38% milling (UVAM & MQL), and 19% drilling (UVAD & MQL) operations. It is worth mentioning that the majority of research done so far has focused on enhancing sustainable and suitable machining techniques for aerospace alloys. The MQL method alone is not effective in machining challenging materials, such as aerospace alloys, due to the phenomenon of effective flow rate. To address this issue, researchers have proposed the use of ultrasonic vibration-assisted turning to improve the effective flow rate in the machining process [30]. When the lubricant

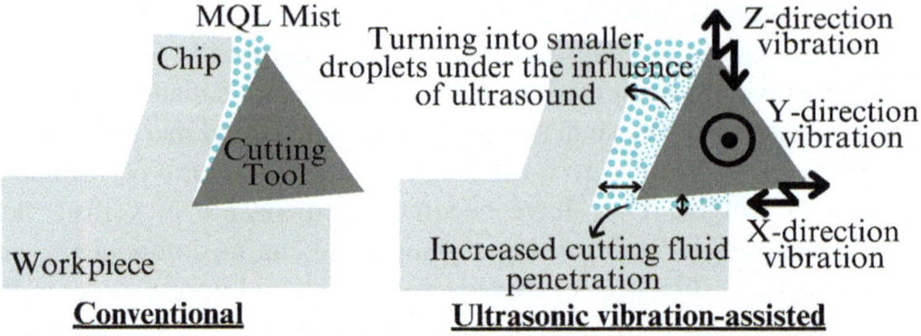

Fig. 3.2 The synergistic effect in UVA machining and MQL technology together

flow rate exceeds its optimal effectiveness threshold, the ratio of air to oil in the mixture ceases to influence both the temperature generated during cutting and the force required to make the cut [31]. Nonetheless, Ji et al. [31] discovered that the optimal flow rate of lubricant in MQL machining processes is dependent on the cutting speed.

This synergistic effect is illustrated in Fig. 3.2 through a schematic diagram. In conventional cutting, the cutting tool and workpiece interface are in continuous contact [32], so it is difficult for MQL droplets to enter the cutting zone. However, in UVA machining with MQL, separation occurs between the tool and the workpiece at a certain frequency and amplitude, facilitating the penetration of MQL droplets into the cutting zone [33, 34]. Furthermore, once the MQL droplets make contact with the tool surface, they will be transformed into smaller and more evenly distributed droplets due to the impact of ultrasound, and this transformation enhances the ability of the MQL droplets to leak into the cutting zone. All these physical mechanisms offer the advantages of the synergistic effect of using UVA machining and MQL method together.

In this chapter, the machining operations of UVA turning, UVA milling, and UVA drilling with MQL are investigated. The process parameters of the investigations, in which UVAT & MQL, UVAM & MQL, and UVAD & MQL methods were studied, in the machining of aerospace alloys are given in Table 3.2.

3.2 Effects on Cutting Forces

Cutting forces are a critical parameter in the machining of aerospace alloys, and the importance of cutting force in machining aerospace alloys lies in its influence on tool life, surface finish, machine stability, energy consumption, workpiece integrity, and overall process stability [57]. Manufacturers in the aerospace industry need to carefully manage cutting forces to achieve the desired quality, precision, and efficiency in the machining of materials used in aerospace applications. High cutting forces can cause machine vibrations and affect the stability of the machining process, and the vibrations can lead to poor surface finish, increased tool wear, and reduced

Table 3.2 Process parameters in UVA machining of aerospace alloys with MQL method

Study	Process	Material	Machining parameters	Vibration parameters	MQL parameters
Agrawal et al. [35]	Turning	Ti-6Al-4V	Environment: CT & dry, CT & MQL, CT & CryoMQL, UVAT & dry, UVAT & MQL, and UVAT CryoMQL Cutting speed: 23, 46, 69 m/min Feed: 0.1, 0.2, 0.3 mm/rev Depth of cut: 0.5 mm	Vibration statement: One-dimensional (1D) and applied tangential direction to the cutting tool Frequency: 20 kHz Amplitude: 10 μm	Cutting fluid: Flow rate: 20 mL/h Pressure: 0.5 MPa Number of nozzle and applied direction: Only rake face of cutting tool Nozzle distance: 20 mm Nozzle angle: 75°
Airao et al. [36]	Turning	Inconel 718	Environment: Dry, flood, MQL, and liquefied carbon dioxide (LCO_2) Cutting speed: 66 m/min Feed: 0.22 mm/rev Depth of cut: 0.3 mm	Vibration statement: 1D and applied tangential direction to the cutting tool Frequency: 20 kHz Amplitude: 10 μm	Cutting fluid: Canola oil Flow rate: 8 mL/h Pressure: 5 bar Number of nozzle and applied direction: Only rake face of cutting tool Nozzle diameter: 2 mm Nozzle distance: 20 mm Nozzle angle: unspecified
Airao et al. [37]	Turning	Ti-6Al-4V	Environment: Dry, flood, MQL, and LCO_2 Cutting speed: 66 m/min Feed: 0.22 mm/rev Depth of cut: 0.3 mm	Vibration statement: 1D and applied tangential direction to the cutting tool Frequency: 20 kHz Amplitude: 10 μm	Cutting fluid: Canola oil Flow rate: 7–8 mL/h Pressure: 5 bar Number of nozzle and applied direction: Only rake face of cutting tool Nozzle diameter: 2 mm Nozzle distance: 20 mm Nozzle angle: Unspecified

(continued)

Table 3.2 (continued)

Study	Process	Material	Machining parameters	Vibration parameters	MQL parameters
Khanna et al. [38]	Turning	Inconel 718	Environment: UVA-MQL, UVA-EMQL, and UVA-LCO_2 Cutting speed: 66 m/min Feed: 0.22 mm/rev Depth of cut: 0.3 mm	Vibration statement: 1D and applied tangential direction to the cutting tool Frequency: 20 kHz Amplitude: 10 μm	Cutting fluid: Canola oil Flow rate: 9 mL/h Pressure: 6 bar Number of nozzle and applied direction: rake and flank face of cutting tool Nozzle diameter: 2 mm Nozzle distance: 20 mm
Yan et al. [39]	Turning	Ti-6Al-4V	Environment: Dry, UVA-Dry, and UVA-MQL Cutting speed: 17.6, 35.2, 52.8, 70.4 m/min Feed: 0.15 mm/rev Depth of cut: 0.75 mm	Vibration statement: Two-dimensional (2D) elliptical and applied radial direction to cutting tool Frequency: 28 kHz Amplitude: 10 μm	Cutting fluid: Vegetable oil Flow rate: 5 mL/h Pressure: unspecified Number of nozzle and applied direction: Only rake face of cutting tool Nozzle diameter: unspecified Nozzle distance: unspecified Nozzle angle: unspecified
Umasekar et al. [40]	Turning	Mg-SiC metal matrix composite (MMC)	Environment: Dry, MQL, UVA-dry, and UVA-MQL Cutting speed: 20, 36, 52 m/min Feed: 0.05, 0.08, 0.11 mm/rev Depth of cut: 0.05, 0.1, 0.15 mm	Vibration statement: 1D and applied tangential direction to cutting tool Frequency: 20 kHz Amplitude: 10 μm	Cutting fluid: Sunflower oil Flow rate: 150 mL/h Pressure: 6 bar Number of nozzle and applied direction: Only rake face of cutting tool Nozzle diameter: unspecified Nozzle distance:

(continued)

Table 3.2 (continued)

Study	Process	Material	Machining parameters	Vibration parameters	MQL parameters
					unspecified Nozzle angle: Unspecified
Liu et al. [41]	Turning	AA-6061 Alloy	Environment: Dry, MQL, NGpn-MQL, ND-MQL UVA-dry, UVA-MQL, NGpn-UVAMQL, and ND-UVAMQL Cutting speed: 20 m/min Feed: 0.1 mm/rev Depth of cut: 0.1 mm	Vibration statement: 2D elliptical vibration and applied tangential and radial directions to the cutting tool Frequency: 40.8 kHz Amplitude: 3 μm to tangential direction and 1 μm to radial direction	Cutting fluid: Palm oil Nano additive: Nanographene sheet, and nanodiamond particle Mass ratio of nanoadditive: 0.5% nanographene and Flow rate: 60 mL/h Pressure: 0.7 MPa Number of nozzle and applied direction: Only rake face of cutting tool Nozzle diameter and distance: unspecified
Alemayehu et al. [42]	Turning	Inconel 718	Environment: CT-MQL and UVA-MQL Cutting speed: 15, 20, 25 m/min Feed: 0.07, 0,13 mm/rev Depth of cut: 0.5 mm	Vibration statement: 1D vibration and applied tangential direction to the cutting tool Frequency: 22.3 kHz Amplitude: 12 μm	Cutting fluid: Sunflower oil Flow rate: 250 mL/h Pressure: 3 bar Number of nozzle and applied direction: Only rake face of cutting tool Nozzle diameter: 2 mm Nozzle distance: unspecified Nozzle angle: unspecified

(continued)

Table 3.2 (continued)

Study	Process	Material	Machining parameters	Vibration parameters	MQL parameters
Lin et al. [43]	Turning	Inconel 718	Environment: Dry, UVA-dry, and UVA-MQL Cutting speed: 25 m/min Feed: 0.05 mm/rev Depth of cut: 0.5 mm	Vibration statement: 2D elliptical vibration and applied tangential and radial directions to the cutting tool Frequency: 40 kHz Amplitude: 7 μm to tangential direction and 7.5 μm to feed direction	Cutting fluid: Palm oil Flow rate: 50 mL/h Pressure: unspecified Number of nozzle and applied direction: Only rake face of cutting tool Nozzle diameter: unspecified Nozzle distance: unspecified
Ni et al. [44]	End-milling	Ti-6Al-4V	Environment: Dry and MQL Spindle speed: 600 rpm Feed per tooth: 8 μm/z Depth of cut: 0.2 mm	Direction: x-axis Type: Longitudinal Vibration applied to the workpiece Frequency: 20 kHz Amplitude: 8 μm	Cutting fluid: Vegetable oil Flow rate: 20 mL/h Pressure: 0.7 MPa Number of nozzles: 2 Nozzle distance: 10 mm Nozzle angle: 45°
Helmy et al. [45]	Edge-trimming	CFRP	Environment: MQL with vortex tube Cutting speed: 200, 300, and 400 m/min Feed rate: 50, 100, and 150 mm/min Depth of cut: 0.25, 0.5, and 1 mm	Direction: z-axis Type: Longitudinal Vibration applied to the cutting tool Frequency: 24.1 kHz Amplitude: Unspecified	Cutting fluid: Blaser Swisslube B-cool 755 water-soluble cutting fluid (%9) Flow rate: 37 mL/min Pressure: 3 bar Number of nozzles: 1 Nozzle distance: unspecified Nozzle angle: unspecified
Ni and Zhu [46]	End-milling	TC4	Environment: Dry and MQL Spindle speed:	Direction: x-axis Vibration applied to the workpiece	Cutting fluid: Vegetable oil Flow rate:

(continued)

Table 3.2 (continued)

Study	Process	Material	Machining parameters	Vibration parameters	MQL parameters
			600, 1200, 1800, and 2400 rpm Feed per tooth: 8 μm/z Depth of cut: 0.2 mm	Type: Longitudinal Frequency: 20 kHz Amplitude: 6, 8, 10, and 12 μm	20 mL/h Pressure: 0.7 MPa Number of nozzles: 2 Nozzle distance: 10 mm Nozzle angle: 45°
Namlu et al. [47]	End-milling	Ti-6Al-4V	Environment: Dry, wet, MQL Spindle speed: 1500, 2000, and 2500 rpm Feed: 0.03, 0.04, 0.05, and 0.06 mm/z Depth of cut: 0.3 and 3 mm	System: Altrasonic Direction: z-axis Vibration applied to the cutting tool Type: Longitudinal Frequency: 19 kHz Amplitude: 10 μm	System: Bielomatik® B1-210 Cutting fluid: Samnos ZM-22 W Flow rate: 50 mL/h Pressure: 5–10 bar Number of nozzles: 1 Nozzle distance: 10 mm Nozzle angle: 30° Nozzle diameter: 5 mm
Namlu et al. [48]	End-milling	Al6061-T6	Environment: Wet and MQL Cutting speed: 150, 200, and 250 m/min Feed: 500, 600, and 700 mm/min Depth of cut: 0.5 mm	System: Altrasonic Direction: z-axis Vibration applied to the cutting tool Type: Longitudinal Frequency: 19547 Hz Amplitude: 10 μm	System: Bielomatik® B1-210 Cutting fluid: Samnos ZM-22 W Flow rate: 50 mL/h Pressure: 8 bar Number of nozzles: 1 Nozzle distance: 10 mm Nozzle angle: 30° Nozzle diameter: 5 mm
Blasl et al. [49]	End-milling	Ti-6Al-4V	Environment: Wet and MQL Cutting speed: 240 m/min for TiAlN coating	System: DMG MORI ULTRASONIC 40 eVo Direction: z-axis Vibration applied	Cutting fluid: Mineral-based oil without additives for TiAlN coating

(continued)

Table 3.2 (continued)

Study	Process	Material	Machining parameters	Vibration parameters	MQL parameters
			carbide tool and 125 m/min for TiAl-based coating carbide tool Feed per tooth: 0.05 μm/z for TiAlN coating carbide tool and 0.04 μm/z for TiAl-based coating carbide tool Depth of cut: 5 mm	to the cutting tool Type: Longitudinal Frequency: 21 kHz Amplitude: 3 μm	carbide tool and alcohol-based oil with additives for TiAl-based coating carbide tool Flow rate: 0.8 mL/min Number of nozzles: 1 Nozzle distance: 21 mm Nozzle angle: 25°
Niu et al. [50]	End-milling	Ti-6Al-4V	Environment: Dry and MQL Cutting speed: 20, 40, 60, 80, and 100 m/min Feed per tooth: 0.04 mm/z Depth of cut: 4 mm Width of cut: 1 mm	System: CKN-T2000 Direction: z-axis Vibration applied to the cutting tool Type: Longitudinal Frequency: 32.6 kHz Amplitude: 5 μm	System: AOOW-NT2 Cutting fluid: Antifreeze water-based cutting fluids Flow rate: 92 mL/h Pressure: 7 bar Number of nozzles: 2 Nozzle distance: 20 mm Nozzle angle: 90° and 70° to feed direction
Wang et al. [51]	Helical milling	CFRP/Ti-6Al-4V stacks	Environment: Dry, MQL, cryogenic (liquid nitrogen [LN_2]), and hybrid (MQL+ LN_2) Spindle speed: 4000 r/min Axial feed: 0.15 mm/r Tangential feed per tooth: 0.033 mm	Direction: z-axis Vibration applied to the cutting tool Type: Longitudinal Frequency: 33 kHz Amplitude: 4–7 μm	System: KS-2106 Cutting fluid: Synthetic lubricating oil, with an anti-wear agent Flow rate: 20 mL/h Pressure: 4.5 bar Number of nozzles: 2 Nozzle distance: 20 mm Nozzle angle: 90° and 70° to feed direction

(continued)

Table 3.2 (continued)

Study	Process	Material	Machining parameters	Vibration parameters	MQL parameters
Zou et al. [52]	Helical milling	Ti-6Al-4V	Environment: Dry, air, and MQL Workpiece hardness: 280 and 350 HV Spindle speed: 2000, 3000, and 4000 rpm Axial feed: 0.1, 0.2, and 0.3 mm/rev Tangential feed per tooth: 0.02, 0.04, and 0.06 mm/tooth	Vibration applied to the cutting tool Type: Longitudinal + torsional Frequency: 33 kHz Amplitude: 4–7 μm	System: unspecified Cutting fluid: Vegetable-based oil Flow rate: 10 mL/h Pressure: 5 bar Number of nozzles: 2 Nozzle distance: 20 mm Nozzle angle: 45°
Isobe and Hara [53]	Drilling	Inconel 600	Environment: Wet and MQL Spindle speed: 3000 rpm Feed rate: 6 mm/min Chip load: 2 μm/rev Step feed: 0.1 mm Hole depth: 2 mm	Vibration applied to the cutting tool Frequency: 40 kHz Amplitude: 2 μm	External MQL System: Fuji BC engineering Cutting fluid: Water-insoluble cutting fluid Number of nozzles: 1 Flow rate: 1.7 ml/min Pressure: 2 bar
Hoang et al. [54]	Drilling	SS304	Environment: Wet, MQL, and Nanofluid MQL Cutting speed: 300 and 500 rpm Feed: 0.05, 0.07, and 0.09 mm/rev Hole depth: 40 mm	Vibration applied to the cutting tool MPI WG-3000 ultrasonic generator Herrmann KHS20-IP50-L transducer Frequency: 20 kHz	Internal MQL AQUA drill ex oil-Hole8D with 5-mm diameter Cutting fluid: Soybean oil for MQL and nanographene-added soybean oil for nanofluid MQL Flow rate: 90 mL/h Pressure: 1.5 bar
Namlu et al. [55]	Drilling	Ni55.8Ti	Environment: Wet and MQL Cutting speed: 15 m/min	Vibration applied to the cutting tool System: Altrasonic	External MQL Internal MQL system System:

(continued)

Table 3.2 (continued)

Study	Process	Material	Machining parameters	Vibration parameters	MQL parameters
			Feed: 100 mm/min Hole depth: 10 mm	Frequency: 18 kHz Amplitude: 8 μm	BielomatikVR B1–210 model Cutting fluid: Vegetable-based oil Number of nozzles: 3 Flow rate: 50 mL/h Pressure: 6 bar Nozzle distance: 10 mm Nozzle angle: 45°
Erturun et al. [56]	Drilling	Inconel 718	Environment: Dry, wet, and MQL Cutting speed: 15 m/min Feed: 0.045 mm/rev Hole depth: 10 mm	Vibration applied to the cutting tool System: Altrasonic Frequency: 18.5 kHz Amplitude: 8 μm	External MQL System: BielomatikVR B1-210 model Cutting fluid: Ester-based oil Number of nozzles: 2 Flow rate: 50 mL/h Pressure: 6 bar Nozzle distance: 40 mm Nozzle angle: 45°

accuracy [58]. In aerospace, where precision is paramount, maintaining stability is critical to meeting tight tolerances. In addition, cutting force has an impact on the energy consumption of the machining process, and higher cutting forces require more energy to overcome resistance, leading to increased power consumption [59]. Reducing cutting forces can contribute to energy efficiency, which is increasingly important in modern manufacturing. Excessive cutting forces can cause damage and poor surface quality, with issues such as surface roughness and chatter marks on the workpiece [60]. For these reasons, in the aerospace industry, where structural integrity is paramount, minimizing the impact of cutting forces on the workpiece is crucial to ensuring that the final component meets safety and performance requirements.

Titanium- and nickel-based alloys are defined as aerospace alloys [61, 62]. However, turning these alloys can be difficult due to their high mechanical properties and low thermal conductivity [62, 63]. During the machining process, these special properties lead to high cutting forces and cutting temperature that leads to tool wear and poor surface roughness [64, 65]. To address these issues, the UVAT & MQL method has been recommended from a sustainability machining standpoint (Table 3.2). Several studies have been conducted to test the effectiveness of this method. Yan et al. [39] used the UVAT & Dry and UVAT & MQL methods in 2D elliptical ultrasonic vibration-assisted turning of Ti-6Al-4V alloy. They found 21.4% and 30% reduction in cutting forces, respectively, compared with the dry conventional turning (CT & Dry) method. As cutting speed increased, cutting forces tended to decrease due to thermal softening. On the other hand, Alemayehu et al. [42] tested the method on Inconel 718 alloy and found that the main cutting forces decreased by 12% when ultrasonic vibration was applied in the tangential direction using the UVAM & MQL method. Similarly, Lin et al. [43] used the UVAM & MQL method to turn Inconel 718 alloy and found that cutting forces were reduced by 61.7% in the tangential direction and 79.7% in the radial direction compared with the CT & Dry method. Although the studies had similar cutting parameters and workpieces, differences in ultrasonic vibration (1D or 2D), frequency and amplitude values, and MQL flow rate affected the results. The studies also did not consider the impact of horn design in ultrasonic systems, tool clamping length, and the natural frequency of the system on performance. Lin et al. [43] also reported that, compared with CT & Dry cutting, temperature reduction is about 10.2% in UVAT & Dry and 16.9% in UVAT & MQL.

It has been determined in various studies that the UVAM & MQL and UVAD & MQL methods reduce the cutting forces when appropriate cutting and vibration parameters are selected (Table 3.2). It is known that, as the vibration amplitude increases in UVAM, the maximum separation distance between the tool and the workpiece increases, the tool–workpiece contact ratio and the actual cutting time decrease, and the cutting force decreases [44]. However, when the amplitude is greater than it should be, it creates a hammer effect and causes an increase in the cutting force [50]. If the effect of cutting speed is evaluated, when the cutting speed is low, UVAM shows good separation properties, reducing the net cutting time, which reduces the cutting force [66]. However, as the cutting speed increases, the cutting mechanism switches from interrupted cutting to continuous impact cutting [46]. Continuous impact cutting, depending on the cutting parameters, can increase the cutting force by increasing the strain hardening in the machining of Ti-6Al-4V alloy or reduce the cutting forces due to the thermal softening due to an increase in temperature [50]. Another reason why the UVAM & MQL method reduces the cutting force even more than UVAM is that the droplets sprayed in MQL become thinner due to the ultrasonic atomization effect as a result of contacting the tool surface under ultrasonic vibration, and this increases the penetration to the cutting zone and reduces the friction force [46]. In the study conducted by Namlu et al. [47], in the rough machining of Ti-6Al-4V alloy, the resultant force of the UVAM & MQL method decreased by 25, 14, and 20%, respectively, compared with the conventional milling under dry cutting condition (CM & Dry), conventional milling under MQL cutting condition (CM & MQL), and UVAM & Dry. On the other hand, in finishing machining, the UVAM & MQL method reduced the resultant force by 17, 27, 26, and 13%, respectively, compared with the CM & Dry, CM & MQL,

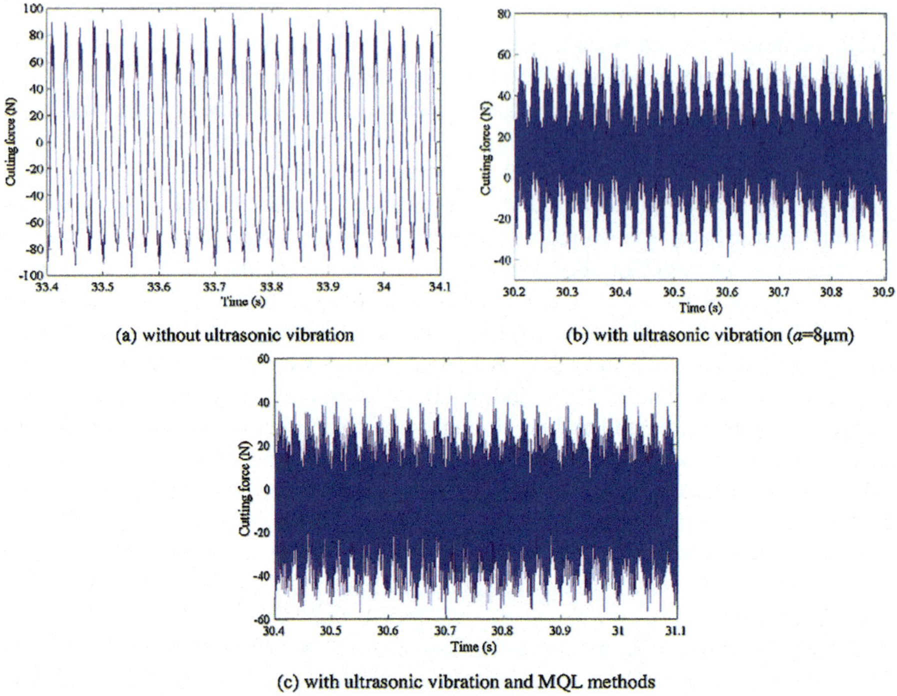

Fig. 3.3 Cutting forces in (**a**) CM, (**b**) UVAM, and (**c**) UVAM & MQL [46]

conventional milling under wet cutting condition (CM & Wet), and ultrasonic vibration-assisted milling under wet cutting condition (UVAM & Wet). Similarly, the cutting forces of the Ti-6Al-4V alloy were investigated by Ni et al. [46] using CM, feed-directional UVAM, and feed-directional UVAM & MQL methods. The cutting force signals of the methods are given in Fig. 3.3, and the results indicate that the average values of cutting force in UVAM & MQL methods are, respectively, reduced by about 30–55% compared with the cutting force in CM. Niu et al. [50] found that using UVAM and UVAM & MQL methods reduced the cutting force by 3.4–18.6% and 26.3–33%, respectively, compared with CM. This reduction in cutting force is attributed to the benefits of intermittent machining, such as decreased actual cutting time and dynamic cutting thickness, along with enhanced lubrication/cooling efficiency, the cavitation effect, and the synergistic effect of these factors.

Erturun et al. [56] obtained an 8.47% improvement in thrust force with the UVAD & MQL method compared with the conventional drilling under the dry cutting condition (CD & Dry) method when drilling holes in Inconel 718 nickel alloy. Aircraft engine parts are mostly made from Inconel 718 nickel alloy, and drilling is an essential operation for these parts as shown in Fig. 3.4 [67]. The improvement in thrust forces is important in meeting this need. Similarly, Isobe et al. [53] achieved a twofold improvement in Inconel 600 alloy. Namlu et al. [55] used the UVAD & MQL method in drilling the nickel–titanium (Ni-Ti) shape memory alloys and

Fig. 3.4 A turbine disk made from Inconel 718 alloy [67]

achieved a 59.6% reduction in thrust force compared with the conventional drilling under the MQL cutting condition (CD & MQL). This improvement is gaining importance due to the increasing utilization of shape memory alloys in the aerospace industry [68].

3.3 Effects on Surface Quality

Surface quality is a critical factor in the manufacturing and performance of aerospace industry parts. It plays a crucial role in ensuring the safety, efficiency, and longevity of aerospace components [69]. For instance, the aerospace sector necessitates extremely precise surface finish specifications for components such as fans and turbine blades. When the surface roughness of the fan blades is high, the risk of debris adhering to the surface of the blades increases, and this increases the engine operating temperature, causing service conditions close to the maximum exhaust temperature band [70]. Thus, the time between the maintenance of aircraft engines is shortened, the need for revision of the parts increases, and they remain in service for a shorter period. Despite all this, if the surface roughness was low, there would be an increase in the time taken for engine maintenance. Moreover, the smooth surface of turbine blades facilitates a more rapid acceleration and compression of air mass flow within the turbines, leading to reduced fuel usage. This efficiency not only conserves energy but also yields economic advantages due to lower operational costs [71]. Reducing surface roughness also contributes to prolonging fatigue life by increasing fatigue resistance [72].

Advanced manufacturing techniques need to be used to reduce surface roughness due to the wide range of benefits offered by low surface roughness in the production of aerospace parts. It has been demonstrated in many scientific studies that UVAT &

Table 3.3 Effect of UVAT & MQL, UVAM & MQL, and UVAD & MQL methods on surface roughness compared with other methods

Study	Material	Machining type	Compared method	Improvement R_a (%)
Airo et al. [37]	Ti-6Al-4V	Turning (UVAT & MQL)	CT & Dry UVAT & Dry CT & MQL	24 26 8
Yan et al. [39]	Ti-6Al-4V	Turning (UVAT & MQL)	CT & Dry UVAT & Dry	33 17
Airo et al. [36]	Inconel 718	Turning (UVAT & MQL)	CT & Dry CT & Wet UVAT & Dry CT & MQL	31 15 25 9
Alemayehu et al. [42]	Inconel 718	Turning (UVAT & MQL)	CT & MQL	27
Liu et al. [41]	AA-6061 Alloy	Turning (UVAT & MQL)	CT & Dry UVAT & Dry CT & MQL	49 52 44
Umasekar et al. [40]	Mg-Sic-MMC	Turning (UVAT & MQL)	CT & Dry UVAT & Dry CT & MQL	21 11 17
Blasl et al. [49]	Ti-6Al-4V	End-milling (UVAM & MQL)	CM & Wet UVAM & Wet CM & MQL	−33 9 −63
Niu et al. [50]	Ti-6Al-4V	End-milling (UVAM & MQL)	CM & Dry UVAM & Dry	50 26
Zou et al. [52]	Ti-6Al-4V	Helical milling (UVAM & MQL)	UVAM & Dry UVAM & Air	46 27
Erturun et al. [56]	Inconel 718	Drilling (UVAD & MQL)	CD & Dry	29

MQL, UVAM & MQL, and UVAD & MQL methods reduce the surface roughness compared with the conventional machining methods for aerospace materials, and the improvement values in surface roughness are given in Table 3.3.

Liu et al. [41] conducted the use of both MQL and UVAT techniques on machining of AA6061. Their results show that the UVAT & MQL lowered the average surface roughness values with a decrease in the adhesion between the tool and the chip, which resulted in fewer chips sticking to the machined surfaces. This is due to the fact that the UVA cutting helped to improve the penetration of cutting fluid in the cutting zone, which in turn reduced friction and led to a decrease in surface roughness. However, the UVAT & Dry process caused regular indentations on the machined surface and increased the surface roughness. Scanning electron microscopy (SEM) analysis indicated that the chips adhere to the machined surfaces in the CT & Dry. Surface topography examinations showed clear plowing marks on the CT & MQL surfaces, and the valley depths were significantly different. The profile curve varied widely from −6.3 μm to 5.7 μm. In contrast, the surface of the UVAT & MQL was much smoother due to good lubrication conditions, which eliminated obvious plowing marks.

However, as can be seen in Table 3.2, the application of UVA machining negatively affected the surface roughness in the study conducted by Blasl et al. [49]. To benefit from the advantages offered by the UVAM & MQL method, there is a critical cutting speed threshold that must be taken into account. If the cutting process is performed above this critical cutting speed, the contact between the tool and the workpiece is not interrupted and remains in constant contact, causing a negative impact on the surface quality [50]. Ultrasonic vibration frequency and amplitude values play an active role in determining the critical cutting speed [44], and therefore, appropriate ultrasonic vibration parameters must be determined. Higher frequencies cause a larger transient angle and more vibrations per cutting arc length, which leads to denser machined surfaces [48]. In comparison, lower frequencies cause an insufficient tool–workpiece intermittent contact mechanism [46]. The amplitude of the ultrasonic vibration generates a separation between the cutting tool and the workpiece. Excessive amplitudes cause a "hammer effect" resulting in undesirable surface damage, whereas if the amplitude of the ultrasonic vibration value is low, it is sufficient to generate a gap for vibration [49]. Namlu et al. [48] studied the surface roughness formation on the AA6061-T6 by changing the ultrasonic vibration frequency and amplitude values at fixed machining parameters and different cutting conditions, and the best surface roughness value was obtained at an amplitude of 10 μm and a frequency of 19 kHz. As can be seen from Fig. 3.5, appropriate values of ultrasonic vibration parameters must be selected to improve surface roughness compared with CM. Similar amplitude and frequency values have been used in the literature in the machining of Ti-6Al-4V alloy using the UVAM & MQL (Table 3.2).

Another effect of the UVAM & MQL method in terms of surface quality is that it provides clean, uniform, and compact microtextured surfaces in surface topography [44, 46–48, 50]. Figure 3.6 shows the surface topographies obtained by Ni et al. [44] as a result of processing the Ti-6Al-4V alloy with different processing methods. In Fig. 3.6a, a large number of machined surface defects (chip adhesion, edge chipping marks, etc.) and the feeding trajectory of the cutting tool are visible on the CM-machined surface. In the UVAM method, the trace of the feeding trajectory of the cutting tool is replaced by the microtextured surface, and it is seen that the

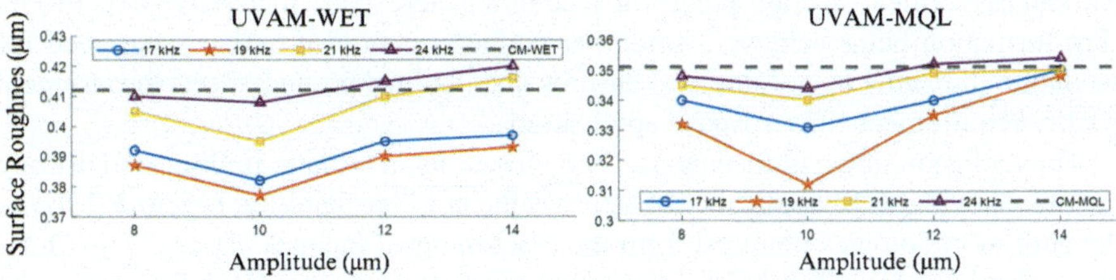

Fig. 3.5 Effect of ultrasonic vibration parameters on surface roughness [48]

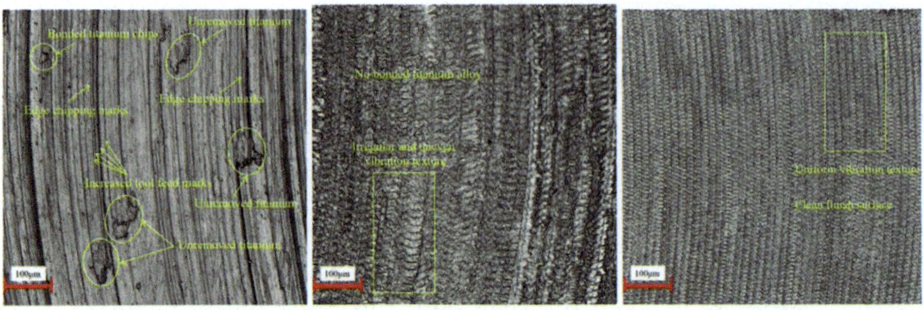

(a) CM (b) UVAM (c) UVAM&MQL

Fig. 3.6 Surface topographies after (**a**) CM, (**b**) UVAM, and (**c**) UVAM & MQL [44]

majority of surface flaws are reduced (Fig. 3.6b). In the UVAM & MQL method, the uniformity and compactness of the microtextured surface in the surface topography increased, as well as a clean texture free from surface defects (Fig. 3.6c).

For aerospace alloys, UVAD & MQL is used to improve the hole surface quality, including surface roughness, roundness, and burr formation [54–56]. Erturun et al. [56] obtained 29.04%, 12.87%, and 26.21% improvements in surface roughness, burr height, and roundness with the UVAD & MQL compared with the CD & Dry method when drilling holes in Inconel 718 nickel alloy. Isobe et al. [53] stated that the UVA & MQL reduces the burr width when drilling holes in Inconel 600 alloy. Namlu et al. [55] determined that UAD & MQL outperformed CD & MQL when it came to drilled hole accuracy, including roundness errors and entry-and-exit hole diameters when drilling Ni55.8Ti shape memory alloy.

3.4 Effects on Chip Morphologies

Chip morphology plays a crucial role in the machining of aerospace alloys, and it has several implications for the overall machining process and the quality of the finished components [57]. The shape and size of chips generated during machining have an impact on tool wear [73]. Proper chip formation can help in reducing tool wear, extending tool life, and minimizing the need for frequent tool changes, which is particularly critical in high-precision machining aerospace materials [74]. Proper chip formation helps achieve a smoother surface finish [75], reduces the need for secondary finishing operations, and ensures that the final product meets the stringent quality requirements in aerospace applications.

The inclusion of the ultrasonic-assisted vibrations in turning, milling, and drilling operations with MQL applications improves the chip breakability. Figure 3.7 shows the chip morphologies obtained from the machining of Inconel 718 and Ti-6Al-4V alloys using the UVAT & MQL. Airao et al. [36] observed that UVAT amplifies the cooling and lubrication effects of MQL during the moment the tool disengages from

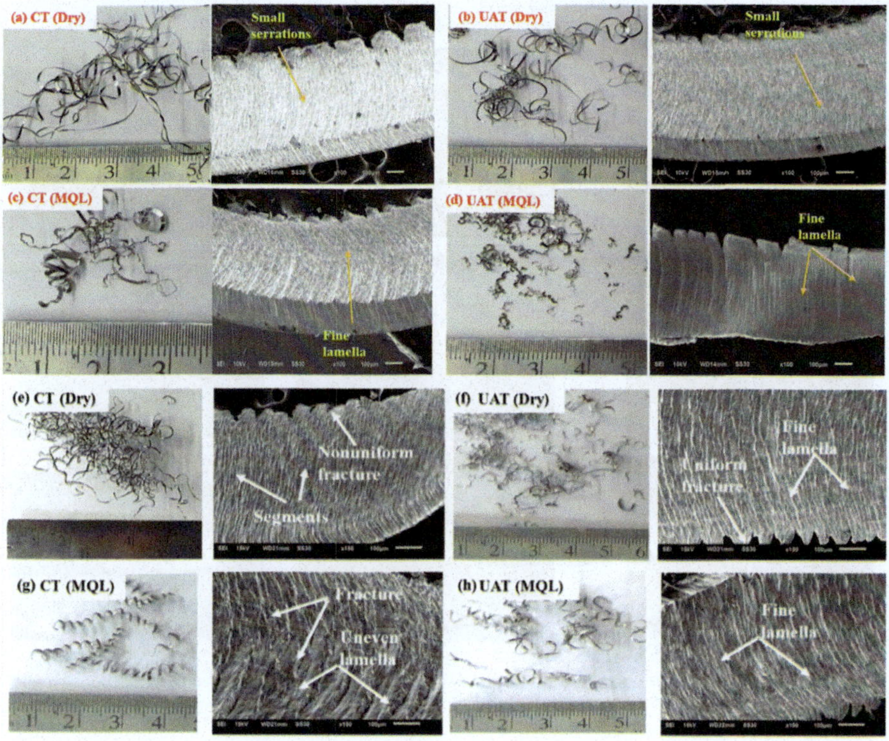

Fig. 3.7 Chip morphologies after (**a**) CT & Dry, (**b**) UVAT & Dry, (**c**) CT & MQL, (**d**) UVAT & MQL for Inconel 718, and (**e**) CT & Dry, (**f**) UVAT & Dry, (**g**) CT & MQL, (**h**) UVAT & MQL for Ti-6Al-4V [36, 37]

the workpiece. Consequently, the chips separate the shear zone, creating distinct sections referred to as semi-continuous chips. These segmented chips are primarily the cause of periodic fluctuations in cutting force and stress levels, which, in turn, lead to a decrease in the lifespan of the tool. The segmentation in the chips could be caused by factors such as plastic deformation instability and physical and metallurgical variations [76]. According to Airo et al. [37], UVAT & MQL simplifies the process of chip removal by enabling a strong airflow to occur between the tool and the chip. In addition, the presence of oil particles reduces friction by creating a lubricating barrier in the contact zone. This makes the process of chip removal much easier and produces short helical chips. Yan et al. [39] reported comparable outcomes on the UVAT & MQL machining of Ti-6Al-4V alloy. They noted that this method generates a desired type of chips, known as the spiral tubular chips.

Although many studies have been published showing that the UVAM method has a positive effect on chip morphology, the investigation of the synergistic effect provided by the UVAM & MQL on chip morphology was limited with the study of Niu et al. [50]. Figure 3.8 shows the effects of CM, UVAM, and UVAM & MQL methods and cutting speeds on chip morphologies in the machining of Ti-6Al-4V.

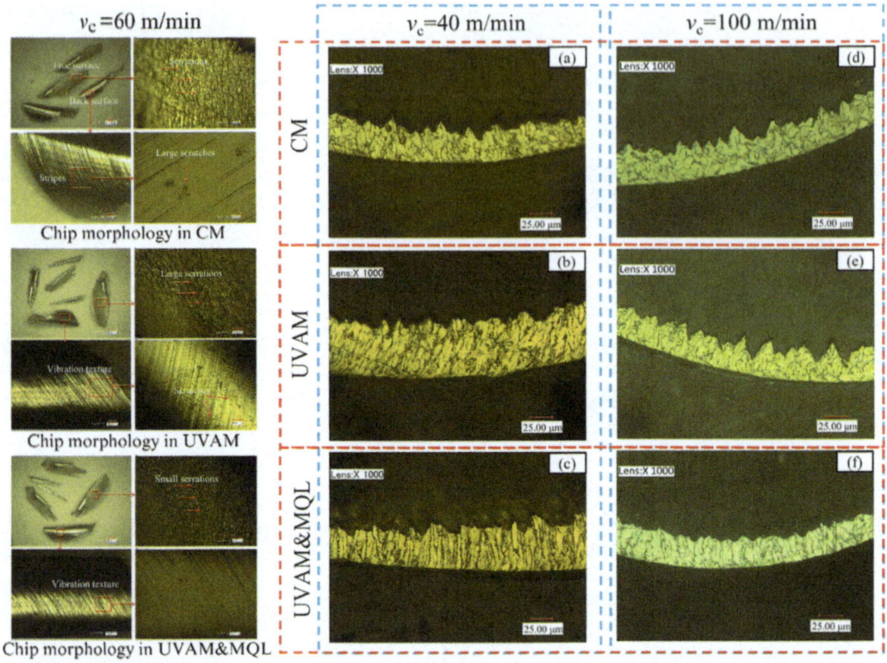

Fig. 3.8 Chip formations in various machining methods [50]

Compared with CM and UVAM, UVAM & MQL has many smaller-sized chips, clearer vibration texture behind the chips, and fewer adhered particles. In addition, decreasing the friction at the tool–chip interface and increasing the heat dissipation in UVAM & MQL behind the chips ensured that the saw tooth height of the saw-type chip was smaller.

Chip breaking in the drilling process facilitates chip removal, increasing tool life and maintaining surface quality, thus providing an improvement in final product quality and cost [77]. The UVAD & MQL method has positive effects on chip flow by increasing chip breaking [54, 56]. When the studies on stainless steel [54] and Inconel 718 [56] were examined, the use of ultrasonic vibration ensured the breaking of chips in both the studies. Using MQL and nanofluid MQL instead of wet cutting, chip sizes become shorter due to the synergistic effect that occurred with the UVAD method.

3.5 Effects on Tool Wear

Flank wear is the most crucial response parameter for assessing the machinability of aerospace alloys and the economics of machining [78, 79]. Titanium alloys are highly reactive at high temperatures and have a chemical affinity with the tool material [80]. As a result, they react chemically with the tool material quickly during

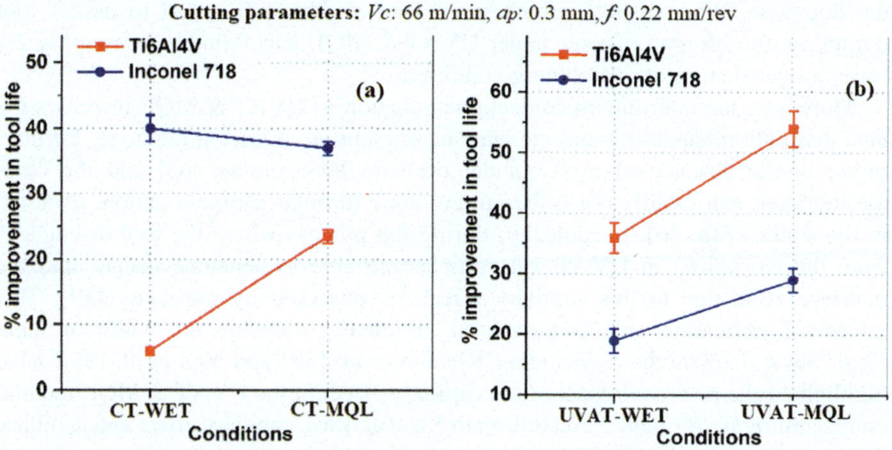

Fig. 3.9 Percentage improvement in tool life for Ti-6Al-4V and Inconel 718 under wet and MQL conditions during (**a**) CT and (**b**) UVAT [84]

machining, leading to chipping and premature tool wear [80, 81]. On the other hand, due to the low elasticity modulus and thermal conductivity, cutting tool wear occurs quickly during the machining of titanium-based and nickel-based alloys such as Inconel containing titanium [82]. In UVA machining, the intermittent cutting process reduces the friction coefficient [83]. Literature reports that UVA machining reduces crater and flank wear, thereby increasing tool life [83]. Besides, when MQL is applied to the UVAT, it causes a synergistic effect on the tool–chip contact area. As a result, the MQL droplets transform into smaller and more evenly distributed droplets due to the effect of ultrasound, enabling them to infiltrate the cutting zone more effectively. Figure 3.9 shows the comparative graphs for tool life and cutting tool flank wear, categorized by material and conditions.

Khanna et al. [84] explored how different cutting conditions impact tool durability during the machining of two popular aviation materials, Ti-6Al-4V and Inconel 718, utilizing the UVAT & MQL technique. Their findings indicated that Inconel 718 had a lower surface tension compared with titanium alloys under conditions of cooling and lubrication. This resulted in a decreased effectiveness of the cutting fluid's lubrication, causing greater wear on tools used with Inconel 718 than those used with Ti-6Al-4V. Notably, Ti-6Al-4V was easier to machine than Inconel 718 under these conditions. The UVAT method allowed for better heat dissipation in the critical zones of tool–chip and tool–workpiece contact, enhancing tool longevity. Both materials experienced less abrasion and tip wear as a result. Specifically, tools used on Ti-6Al-4V lasted about 9% and 34% longer than those on Inconel 718 with CT and UVAT, respectively. The MQL technique, involving a pressurized mix of air and oil particles, further increased tool life by forming a lubrication layer and efficiently cooling the area of deformation, thus prolonging the life of the cutting tool [85]. Yan et al. [39] discovered that applying UVAT & MQL at a cutting speed of 70 m/min notably slowed down the rate of flank wear development. When using a flank wear land width of 0.3 mm to assess tool durability, the lifespan of tools under UVAT & MQL was found to be roughly 2.6 times longer than in dry machining conditions.

Moreover, the intermittent contact phenomenon in UVAT & MQL increases the heat dissipation capability and prevents heat retention at the cutting zone. Furthermore, in the absence of high contact pressure between the tool and the chip, the lubricant can rapidly reach the cutting zone through capillary action, as noted in the studies [38, 86]. Specifically, during the process where the tool disengages from the workpiece in UVAT, oil particles are able to penetrate deeper into the interface zone due to this capillary effect, as observed by the study [39]. This improves lubrication and heat removal, ultimately reducing the width of flank wear. Several researchers, including Khanna et al. [38] and Yan et al. [39], have highlighted the positive impact of the capillary effect in the UVAT & MQL method on machining performance. According to Xu et al. [86], capillary effect and capillary force can be defined in Eq. 3.1.

$$F_{\text{cap}} = \Delta P\, \pi\, r^2 = 2\, \pi\, r\, \gamma\, \cos(\theta) \tag{3.1}$$

In this equation, F_{cap} is the capillary force, ΔP is the driving pressure in the capillary, r is the capillary radius, γ is the lubricant surface tension, and θ is the contact angle. A schematic illustration of the capillary effect is shown in Fig. 3.10.

The UVAM & MQL method has been found to significantly prevent cutting tool flank wear due to the synergistic and capillary effects. In the studies conducted by Airao et al. [37] and Yan et al. [39], Ti-6Al-4V alloy was machined using CT & Dry, UVAT & Dry, and UVAT & MQL methods. In both the studies, it was observed that the CT & Dry resulted in adhesive wear and built-up edge (BUE) mechanism due to high cutting temperatures. However, the cooling effect in the UVAT & MQL cutting eliminated the possibility of diffusion wear caused by a chemical reaction between the cutting tool and Ti-6Al-4V alloy. Figure 3.11 shows the tool wear images after the investigated machining methods.

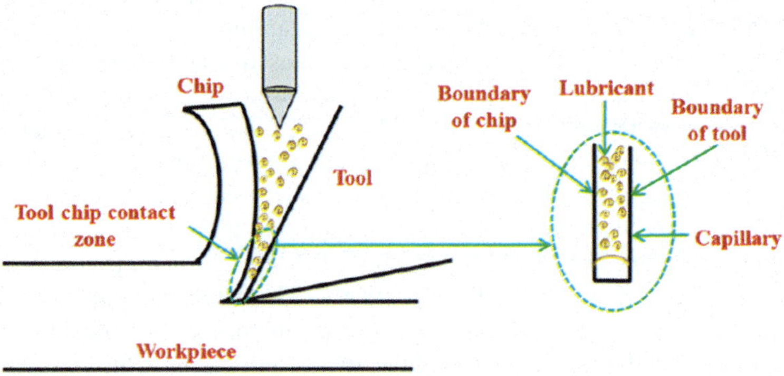

Fig. 3.10 Capillary effect in UVAT & MQL method [38]

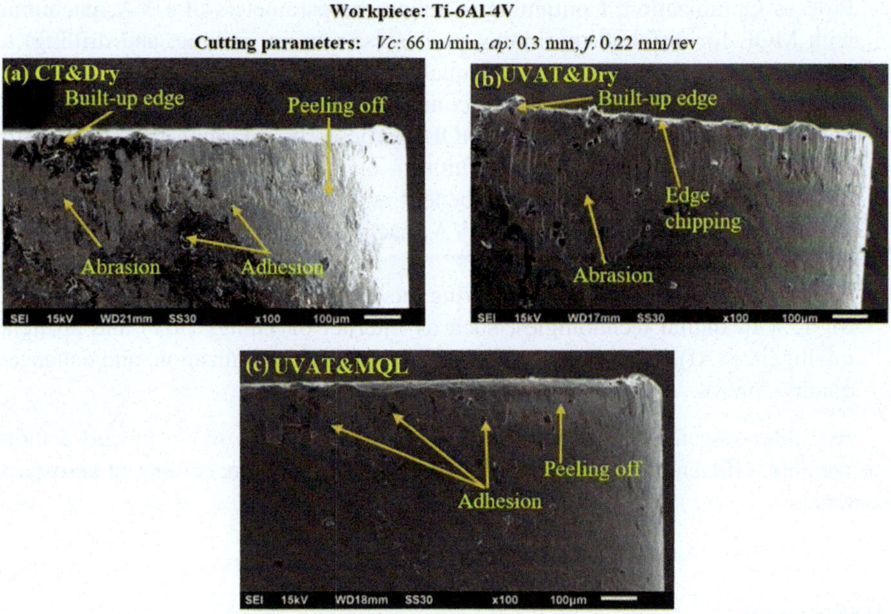

Fig. 3.11 Flank wear images after (**a**) CT & Dry, (**b**) UVAT & Dry, and (**c**) UVAT & MQL [37]

3.6 Summary and Future Scope

In conclusion, this chapter has provided a comprehensive review of UVA machining techniques–specifically turning, milling, and drilling–in combination with MQL, as applied to aerospace materials. The insights gathered from the various sections underscore the significant advancements and benefits these technologies offer.

The effects on cutting forces revealed that UVA machining with MQL can substantially reduce the force required during machining, leading to more efficient operations and less strain on machine components. In terms of surface quality, the application of these techniques has been shown to enhance the finishing and structural integrity of aerospace materials, which is critical for their performance and longevity. The examination of chip morphologies provided valuable insights into the material removal processes, emphasizing the cleaner and more controlled chip formation with UVA machining and MQL. Lastly, the discussion on tool wear highlighted the potential of these methods to extend tool life, thereby reducing costs and downtime in manufacturing processes.

For future work, several areas have been identified where further research and development could be particularly beneficial:

- Material Diversity: Expanding the range of aerospace materials to include newer composites and alloys, ensuring broader applicability of the findings.

- Process Optimization: Continuing to refine the parameters of UVA machining with MQL for different machining processes (turning, milling, and drilling) to achieve even greater efficiency and quality.
- Environmental Impact: Conducting comprehensive studies on the environmental benefits of UVA machining with MQL, particularly regarding the reduction in lubricant usage and energy consumption.
- Economic Analysis: Detailed cost–benefit analyses to assess the economic viability of widespread adoption of UVA machining with MQL in the aerospace industry.
- Integration with Industry 4.0: Exploring the combination of UVA machining and MQL with digital technologies, such as Internet of Things (IoT) and artificial intelligence (AI) for predictive maintenance, process optimization, and enhanced quality control.

By addressing these future research areas, the field can move toward a more sustainable, efficient, and economically viable future in the machining of aerospace materials.

References

1. Madhwal Y, Avdeeva Z. Planning in aircraft industry based on prediction of air traffic. Procedia Comput Sci. 2017;122:1047–54.
2. Saoubi RM, Axinte D, Leung S, Nobel C, Attia H, Kappmeyer G, Engin S, Sim W. High performance cutting of advanced aerospace alloys and composite materials. CIRP Ann Manuf Technol. 2015;64:557–80.
3. Williams JC, Starke EA. Progress in structural materials for aerospace systems. Acta Mater. 2003;51(19):5775–99.
4. Tools SE. Turning difficult-to-machine alloys. Techn Guide. 2002:21.
5. Boyer RR. An overview on the use of titanium in aerospace industry. J Mater Sci Eng. 1996; A213:103–14.
6. Ezugwu EO, Wang ZM. Titanium alloys and their machinability—a review. J Mater Process Technol. 1997;68:262–74.
7. https://www.custompartnet.com/process-selector. Accessed 25 Jan 2023.
8. Kaufman JG. Properties of aluminum alloys: tensile, creep, and fatigue data at high and low temperatures. ASM International; 1999.
9. Polmear IJ. Light alloys: from traditional alloys to nanocrystals. Elsevier; 2006.
10. Davis JR. Alloying: understanding the basics. ASM International; 2001.
11. https://www.matweb.com/. Accessed 25 Jan 2023.
12. Kainer KU, editor. Magnesium alloys and technology. Wiley-VCH; 2003.
13. Kainer KU, editor. Magnesium. In: Proceedings of the 6th international conference magnesium alloys and their applications; 2003.
14. Froes FH, Qian M. Developments in magnesium alloy applications in aerospace and electronic industries. JOM J Miner Met Mater Soc. 2016;
15. Avedesian MM, Baker H, editors. ASM specialty handbook: magnesium and magnesium alloys. ASM International; 1999.
16. Verhoeven JD. Fundamentals of physical metallurgy. Wiley; 1975.
17. Ashby MF. Materials selection in mechanical design. Butterworth-Heinemann; 2011.
18. Llewellyn DT, Hudd RC. Steels: metallurgy and applications. Butterworth-Heinemann; 1998.

19. Roberts G, Kennedy R. Tool Steels. ASM International; 1998.
20. Davis JR, editor. ASM specialty handbook: carbon and alloy steels. ASM International; 1996.
21. Miller S. Advanced materials means advanced engines. Interdiscipl Sci Rev. 1996;21(2): 117–29.
22. Lauwers B, Klocke F, Klink A, Tekkaya AE, Neugebauer R, Mcintosh D. Hybrid processes in manufacturing. CIRP annals—manufacturing. Technology. 2014;63:561–83.
23. Gibson AG, Tzavaras A. Composite materials for aerospace applications. Compos A: Appl Sci Manuf. 2010;41(10)
24. Gibson LJ, Ashby MF. Materials selection in mechanical design. Mater Sci Eng A. 1997;204 (1–2):11–23.
25. M'saoubi R, Axinte R, Soo SL, Nobel C, Attia H, Kappmeyer G, Engin S, Sim WM. High performance cutting of advanced aerospace alloys and composite materials. CIRP Ann. 2015;64:557–80.
26. Chu MQ, Ding RG, Zhang SY, Zheng JP, Zhang N. Surface integrity for machining aerospace parts. Mater Rep. 2021;35(7):7183–9.
27. Ulutan D, Özel T. Machining induced surface integrity in titanium and nickel alloys: a review. Int J Mach Tools Manuf. 2011;51(3):250–80.
28. Esmaeili H, Adibi H, Rezaei SM. Study on surface integrity and material removal mechanism in eco-friendly grinding of Inconel 718 using numerical and experimental investigations. Int J Adv Manuf Technol. 2021;112:1797–818.
29. Żyra A, Bogucki R, Podolak-Lejtas A, Skoczypiec S. Research on influence of heat treatment scheme of Ti10V2Fe3Al alloy on technological surface integrity after electrodischarge machining. J Manuf Process. 2021;62:47–57.
30. Wang X, Li C, Zhang Y, Ding W, Yang M, Gao T, et al. Vegetable oil-based nanofluid minimum quantity lubrication turning: academic review and perspectives. J Manuf Process. 2020;59:76–97.
31. Ji X, Li B, Zhang X, Liang SY. The effects of minimum quantity lubrication (MQL) on machining force, temperature, and residual stress. Int J Precis Eng Manuf. 2014;15:2443–51.
32. Xu WX, Zhang LC. Ultrasonic vibration-assisted machining: principle, design and application. Adv Manuf. 2015;3:173–92.
33. Li K-M, Wang S-L. Effect of tool wear in ultrasonic vibration-assisted micro-milling. Proc Inst Mech Eng B J Eng Manuf. 2014;228(6):847–55.
34. Rasidi I, Rahim EA, Ibrahim AA, Maskam NA, Ghani SC. The effect on the application of coolant and ultrasonic vibration assisted micro milling on machining performance. AMM. 2014;660:65–9.
35. Agrawal C, Khanna N, Gupta MK, Kaynak Y. Sustainability assessment of in-house developed environment-friendly hybrid techniques for turning Ti-6Al-4V. Sustain Mater Technol. 2020;26:e00220.
36. Airao J, Nirala CK, Khanna N. Novel use of ultrasonic-assisted turning in conjunction with cryogenic and lubrication techniques to analyze the machinability of Inconel 718. J Manuf Process. 2022;81:962–75.
37. Airao J, Nirala CK, Bertolini R, Krolczyk GM, Khanna N. Sustainable cooling strategies to reduce tool wear, power consumption and surface roughness during ultrasonic assisted turning of Ti-6Al-4V. Tribol Int. 2022;169:107494.
38. Khanna N, Airao J, Nirala CK, Krolczyk GM. Novel sustainable cryo-lubrication strategies for reducing tool wear during ultrasonic-assisted turning of Inconel 718. Tribol Int. 2022;174: 107728.
39. Yan L, Zhang Q, Yu J. Effects of continuous minimum quantity lubrication with ultrasonic vibration in turning of titanium alloy. Int J Adv Manuf Technol. 2018;98:827–37.
40. Umasekar V, John MS. Experimental investigations on ultrasonic assisted turning of magnesium silicon carbide. Mater Manuf Process. 2023;39:1–11.

41. Liu G, Wang J, Zheng J, Ji M, Wang X. An experimental study on ultrasonic vibration-assisted turning of aluminum alloy 6061 with vegetable oil-based Nanofluid minimum quantity lubrication. Lubricants. 2023;11(11):470.
42. Alemayehu H, Ghosh S, Rao P, editor. Evaluation of cutting force and surface roughness of inconel 718 using a hybrid ultrasonic vibration-assisted turning and minimum quantity lubrication (MQL). In: Advances in unconventional machining and composites: proceedings of AIMTDR. Springer; 2020.
43. Lin J, Wang D, editor. Effect of different nozzle angles on turning Inconel718 based on UEV. In: 3rd international conference on electronic information technology and computer engineering (EITCE). IEEE; 2019.
44. Ni C, Zhu L, Yang Z. Comparative investigation of tool wear mechanism and corresponding machined surface characterization in feed-direction ultrasonic vibration assisted milling of Ti–6Al–4V from dynamic view. Wear. 2019;436-437:203006.
45. Helmy MO, El-Hofy MH, El-Hofy H. Influence of process parameters on the ultrasonic assisted edge trimming of aerospace CFRP laminates using MQL. Int J Mach Mach Mater. 2020;22: 349–73.
46. Ni C, Zhu L. Investigation on machining characteristics of TC4 alloy by simultaneous application of ultrasonic vibration assisted milling (UVAM) and economical-environmental MQL technology. J Mater Process Technol. 2020;278:116518.
47. Namlu RH, Sadigh BL, Kiliç SE. An experimental investigation on the effects of combined application of ultrasonic assisted milling (UAM) and minimum quantity lubrication (MQL) on cutting forces and surface roughness of Ti-6AL-4V. Mach Sci Technol. 2021;25(5):738–75.
48. Namlu RH, Yılmaz OD, Sadigh BL, Kılıç SE. An experimental study on surface quality of Al6061-T6 in ultrasonic vibration-assisted milling with minimum quantity lubrication. Procedia CIRP. 2022;108:311–6.
49. Blasl J, Lichtinger K, Vieltorf F, et al. Experimental investigation of ultrasonic vibration-assisted cryogenic minimum quantity lubrication for milling of Ti-6Al-4V and grinding of Zerodur. Prod Eng Res Devel. 2024;18:75–86.
50. Niu Q, Rong J, Jing L, Gao H, Tang S, Qiu X, Liu L, Wang X, Dai F. Study on force-thermal characteristics and cutting performance of titanium alloy milled by ultrasonic vibration and minimum quantity lubrication. J Manuf Process. 2023;95:115–30.
51. Wang J, Ge J, Chen G, Liu J, Wang Z, Ren C. Sustainable cooling/lubrication induced thermo-mechanical effects on ultrasonic vibration helical milling of CFRP/Ti–6Al–4V stacks. Int J Lightweight Mater Manuf. 2023;6(3):311–28.
52. Zou Y, Guo S, Li H, et al. Investigation on cutting performance in ultrasonic assisted helical milling of Ti6Al4V alloy by various parameters and cooling strategies. Int J Adv Manuf Technol. 2023;126:5123–38.
53. Isobe H, Hara K. Improvement of drill life for nickel super alloy by ultrasonic vibration machining with minimum quantity lubrication. KEM. 2014;625:581–6.
54. Hoang T-D, Ngo Q-H, Chu N-H, Mai T-H, Nguyen T, Ho K-T, Nguyen D. Ultrasonic assisted nano-fluid MQL in deep drilling of hard-to-cut materials. Mater Manuf Process. 2022;37(6): 712–21.
55. Namlu RH, Lotfi B, Kılıç SE, Yılmaz OD, Akar S. Combined use of ultrasonic-assisted drilling and minimum quantity lubrication for drilling of NiTi shape memory alloy. Mach Sci Technol. 2023;27(4):325–49.
56. Erturun ÖF, Tekaüt H, Çiçek A, et al. An experimental study on ultrasonic-assisted drilling of Inconel 718 under different cooling/lubrication conditions. Int J Adv Manuf Technol. 2024;130: 665–82.
57. Ezugwu EO. Key improvements in the machining of difficult-to-cut aerospace superalloys. Int J Mach Tools Manuf. 2005;45(12–13):1353–67.
58. Chuangwen X, Jianming D, Yuzhen C, Huaiyuan L, Zhicheng S, Jing X. The relationships between cutting parameters, tool wear, cutting force and vibration. Adv Mech Eng. 2018;10(1): 1–14.

59. Pimenov DY, Mia M, Gupta MK, Machado ÁR, Pintaude G, Unune DR, Khanna N, Khan AM, Tomaz Í, Wojciechowski S, Kuntoğlu M. Resource saving by optimization and machining environments for sustainable manufacturing: a review and future prospects. Renew Sust Energ Rev. 2022;166:112660.
60. Bolar G, Das A, Joshi SN. Measurement and analysis of cutting force and product surface quality during end-milling of thin-wall components. Measurement. 2018;121:190–204.
61. Tayal A, Kalsi NS, Gupta MK. Machining of superalloys: a review on machining parameters, cutting tools, and cooling methods. Mater Today Proc. 2021;43:1839–49.
62. Yin Q, Liu Z, Wang B. Machinability improvement of Inconel 718 through mechanochemical and heat transfer effects of coated surface-active thermal conductive mediums. J Alloys Compd. 2021;876:160186.
63. Kandi R, Sahoo SK, Sahoo AK. Ultrasonic vibration-assisted turning of titanium alloy Ti–6Al–4V: numerical and experimental investigations. J Braz Soc Mech Sci Eng. 2020;42(8):1–17.
64. Mahesh K, Philip JT, Joshi S, Kuriachen B. Machinability of Inconel 718: a critical review on the impact of cutting temperatures. Mater Manuf Process. 2021;36(7):753–91.
65. Liang X, Liu Z, Wang B. State-of-the-art of surface integrity induced by tool wear effects in machining process of titanium and nickel alloys: a review. Measurement. 2019;132:150–81.
66. Zhu L, Ni C, Yang Z, Liu C. Investigations of micro-textured surface generation mechanism and tribological properties in ultrasonic vibration-assisted milling of Ti–6Al–4V. Precis Eng. 2019;57:229–43.
67. Shlyannikov VN, Zakharov AP, Yarullin RR. Structural integrity assessment of turbine disk on a plastic stress intensity factor basis. Int J Fatigue. 2016;92:234–45.
68. Hartl DJ, Lagoudas DC. Aerospace applications of shape memory alloys. Proc Inst Mech Eng Part G: J Aerosp Eng. 2007;221(4):535–52.
69. Monaca A, Murray JW, Liao Z, Speidel A, Robles-Linares JA, Axinte DA, Hardy MC, Clare AT. Surface integrity in metal machining—part II: functional performance. Int J Mach Tools Manuf. 2021;164:103718.
70. Mediratta R, Ahluwalia K, Yeo SH. State-of-the-art on vibratory finishing in the aviation industry: an industrial and academic perspective. Int J Adv Manuf Technol. 2016;85:415–29.
71. Alqallaf J, Teixeira JA. Blade roughness effects on compressor and engine performance-a CFD and thermodynamic study. Aerospace. 2021;8(11):330.
72. Javidi A, Rieger U, Eichlseder W. The effect of machining on the surface integrity and fatigue life. Int J Fatigue. 2008;30(10–11):2050–5.
73. Zhang G, To S, Xiao G. The relation between chip morphology and tool wear in ultra-precision raster milling. Int J Mach Tools Manuf. 2014;80–81:11–7.
74. Dargusch MS, Sun S, Kim JW, Li T, Trimby P, Cairney J. Effect of tool wear evolution on chip formation during dry machining of Ti-6Al-4V alloy. Int J Mach Tools Manuf. 2018;126:13–7.
75. Zhao Q, Guo X, Wang H, Yin S, Kang R, Dong Z, Bao Y. Effects of ultrasonic vibration cutting trajectories on chip formation of tungsten alloys. J Manuf Process. 2023;92:147–56.
76. Airao J, Kishore H, Nirala CK. Tool wear behavior in μ-turning of nimonic 90 under vegetable oil-based cutting fluid. J Micro Nano Manuf. 2021;9(4):041003.
77. Kumar MS, Deivanathan R. Effect of process parameters on drilling—an overview. Mater Today: Proc. 2021;46(2):1401–6.
78. Singh J, Gill SS, Mahajan A. Experimental investigation and optimizing of turning parameters for machining of Al7075-T6 aerospace alloy for reducing the tool Wear and surface roughness. J Mater Eng Perform. 2023:1–12.
79. Luo H, Zhang D, Luo M. Tool wear and remaining useful life estimation of difficult-to-machine aerospace alloys: a review. China Mech Eng. 2021;32(22):2647.
80. Ezugwu E, Da Silva R, Bonney J, Machado A. Evaluation of the performance of CBN tools when turning Ti–6Al–4V alloy with high pressure coolant supplies. Int J Mach Tools Manuf. 2005;45(9):1009–14.

81. Ezugwu EO, Bonney J, Da Silva RB, Çakir O. Surface integrity of finished turned Ti–6Al–4V alloy with PCD tools using conventional and high pressure coolant supplies. Int J Mach Tools Manuf. 2007;47(6):884–91.
82. Farahnakian M, Keshavarz ME, Elhami S, Razfar MR. Effect of cutting edge modification on the tool flank wear in ultrasonically assisted turning of hardened steel. Proc Inst Mech Eng B J Eng Manuf. 2019;233(5):1472–82.
83. Yang Z, Zhu L, Zhang G, Ni C, Lin B. Review of ultrasonic vibration-assisted machining in advanced materials. Int J Mach Tools Manuf. 2020;156:103594.
84. Khanna N, Airao J, Kshitij G, Nirala CK, Hegab H. Sustainability analysis of new hybrid cooling/lubrication strategies during machining Ti6Al4V and Inconel 718 alloys. Sustain Mater Technol. 2023;36:e00606.
85. Mishra SK, Ghosh S, Aravindan S. Machining performance evaluation of Ti6Al4V alloy with laser textured tools under MQL and nano-MQL environments. J Manuf Process. 2020;53:174–89.
86. Xu X, Lv T, Luan Z, Zhao Y, Wang M, Hu X. Capillary penetration mechanism and oil mist concentration of Al2O3 nanoparticle fluids in electrostatic minimum quantity lubrication (EMQL) milling. Int J Adv Manuf Technol. 2019;104:1937–51.

Chapter 4
Smart Electrical Discharge Machining of Aerospace Alloys

Namadi Vinod Kumar, P. M. Abhilash, and D. Chakradhar

4.1 Introduction

In the ever-evolving field of aerospace engineering, the pursuit of precision and innovation stands as an unwavering directive. Navigating the intricate geometries of aircraft and spacecraft components poses formidable challenges and demands groundbreaking solutions. Traditional electrical discharge machining (EDM), synchronized with artificial intelligence (AI) techniques, emerges as a transformative synergy at the forefront of this pursuit.

Aerospace components often require machining with the utmost precision. The aerospace industry not only involves the selection of high-performance materials but also requires producing intricate geometries, ensuring proper assembly where precision plays a pivotal role. From turbine blade production to aircraft engine components and from structural elements to fuel injection nozzles, precise geometries are paramount for ensuring the stability, reliability, safety, and robustness of aerospace components. To better illustrate the importance of precision, consider the cooling system of the turbine blade, as shown in Fig. 4.1.

In aircraft turbine blades, the temperature may rise more than 1800 K and the blades made from superalloys such as nickel-based alloys need an additional cooling system to perform well in high temperature conditions. Hence, there is a need to drill a large number of holes on each blade, which is possible using EDM process.

Film cooling holes and the tip cap holes are integral features in turbine blade design. They serve to release cooling air and create a protective film. This film acts as

N. V. Kumar · D. Chakradhar
Department of Mechanical Engineering, Indian Institute of Technology Palakkad, Palakkad, Kerala, India

P. M. Abhilash (✉)
Centre for Precision Manufacturing, DMEM, University of Strathclyde, Glasgow, UK
e-mail: abhilash.p-m@strath.ac.uk

S. Gürgen (ed.), *Advances in Manufacturing for Aerospace Alloys*, Sustainable Aviation, https://doi.org/10.1007/978-3-031-64455-9_4

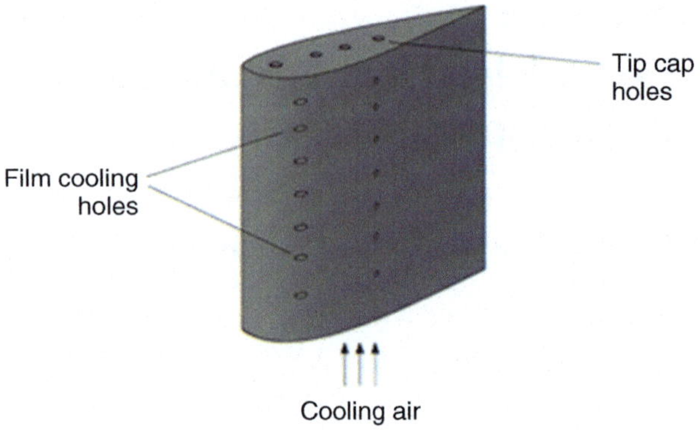

Fig. 4.1 Cooling system of the turbine blade

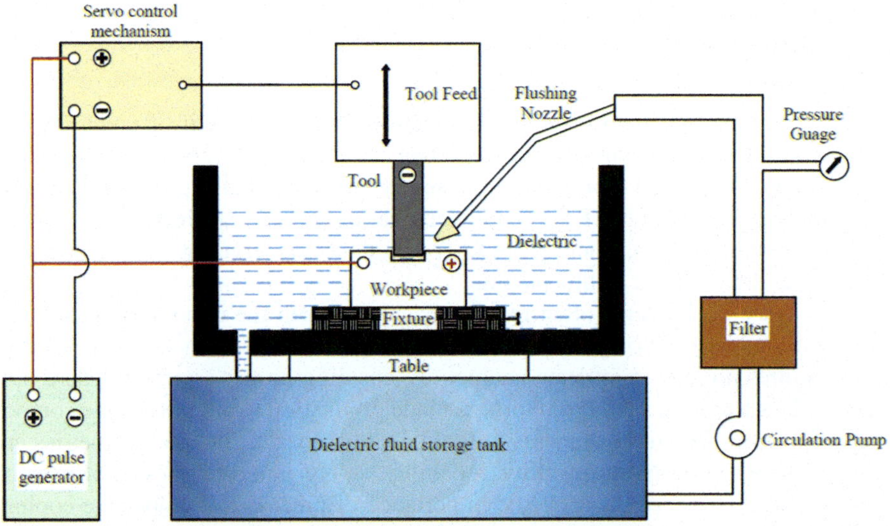

Fig. 4.2 Schematic diagram of the EDM process

a barrier against high temperatures in gas turbines, safeguarding the structural integrity of the blade and contributing to improved overall efficiency without compromising aerodynamics or strength. The efficiency of the EDM process to produce these holes is dependent on key process parameters and various machining conditions to ensure stable machining [1]. These challenges are addressed by the integration of EDM and artificial intelligence techniques, as explained further in the following sections.

EDM is an advanced machining process that has gained popularity in the precision manufacturing industries. The material removal process is governed by the application of a series of electrical discharges known as sparks [2]. As a result, the material is eroded by melting and vaporization. Figure 4.2 illustrates the schematic diagram of the EDM process.

In this process, the sparks generated between the tool and the workpiece are responsible for the removal of the material, resulting in the crater formation in the workpiece material. These small craters collectively constitute the surface roughness of the workpiece. Thus, precise adjustment of process parameters is essential to achieve the desired outcomes, contributing to the development of accurate aerospace components. Smart EDM, driven by AI techniques, facilitates the achievement of these goals, a topic that will be explored further in this chapter. With the ability to produce aerospace components with high accuracy at the micrometer level, EDM aligns with the stringent requirements of aerospace applications. The following points provide an overview of and insights into the extensive scope of EDM in machining aerospace alloys.

1. The production of cooling holes in turbine blades using EDM, resulting in a surface roughness of less than 1 micron [3].
2. Machining of the fir-tree slots of turbine blades.
3. Producing seal slots in turbine vanes.
4. Drilling of holes in fuel injection nozzles.
5. Complex patterns in actuators and motors in aircraft systems.
6. Manufacturing structural aircraft components such as brackets, fittings, and connectors.
7. Aerospace fasteners machining with complex shapes and tight tolerances

Figure 4.3 shows the components that can be machined using EDM.

EDM stands as a pivotal tool in the machining of aerospace alloys, given its significance in handling complex machining processes governed by numerous parameters. Achieving optimal performance in the machining of aerospace components through EDM requires meticulous optimization of various process parameters. Although the application of parametric optimization has shown success in ensuring stable machining processes, the dynamic nature of EDM, coupled with experimental anomalies, calls for a more sophisticated approach, namely the application of AI [4]. Smart EDM, in this context, signifies the evolution of traditional EDM through the seamless integration of artificial intelligence. This advanced iteration harnesses intelligent algorithms, including machine learning (ML) and adaptive control, to dynamically optimize and adapt the machining process.

In response to the evolving challenges inherent in the EDM process, this chapter discusses the paradigm shift from traditional EDM to transformative integration with AI as shown in Fig. 4.4, giving rise to smart EDM. It represents an advanced iteration where AI techniques, such as machine learning and adaptive control, are seamlessly integrated into the EDM process. This incorporation aims to improve precision, stability, and adaptability, ushering in a new era of intelligent machining for aerospace alloys. This section explores the rationale behind incorporating AI, with the aim of redefining the landscape of precision manufacturing in the aerospace industry.

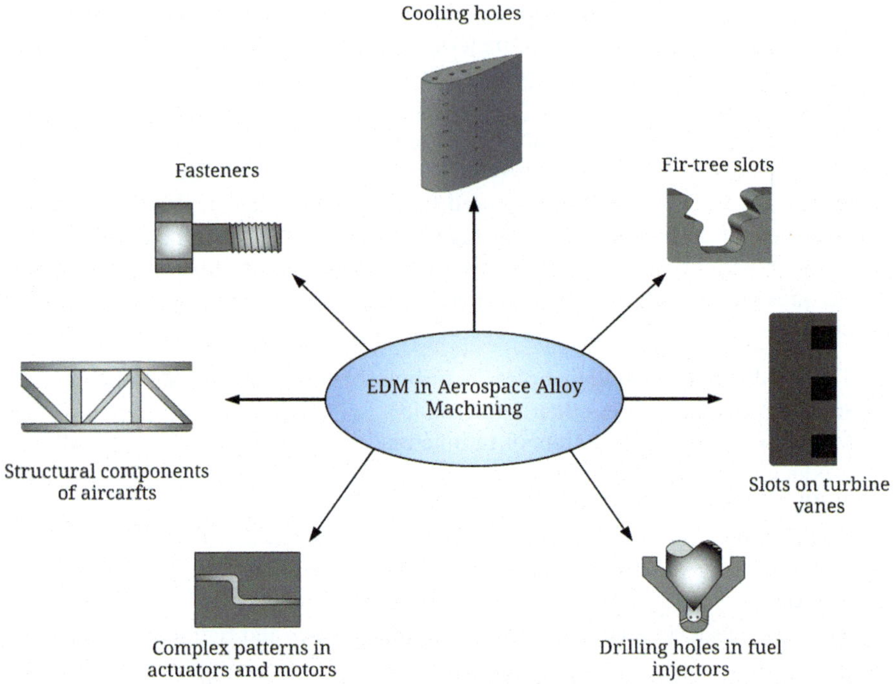

Fig. 4.3 Aerospace components produced with EDM

4.2 Overview on EDM

EDM is a sophisticated nonconventional machining process that relies on controlled electrical discharges to shape materials precisely. In this intricate method, the tool and the workpiece are meticulously brought into close proximity. The voltage is applied between the tool and the workpiece, creating a potential difference. As this potential difference increases, it reaches a critical point, causing a breakdown of the dielectric fluid or gas that surrounds the tool and the workpiece. This breakdown results in the formation of sparks or electrical discharges, igniting a controlled and an intense burst of energy.

The electrical discharges lead to a drop in voltage across the spark gap, creating a pathway for the passage of current. A simple image of a single discharge is shown in Fig. 4.5 for illustrative purposes. The localized high temperature generated by these electrical discharges causes the material in their vicinity to undergo rapid heating, melting, and vaporization. The material removal occurs in small, controlled increments, gradually shaping the workpiece with unparalleled precision.

The dielectric fluid or gas plays a crucial role in the process, not only by facilitating the formation of sparks but also by flushing away the eroded material, ensuring a clean and an efficient machining environment. The entire EDM process is

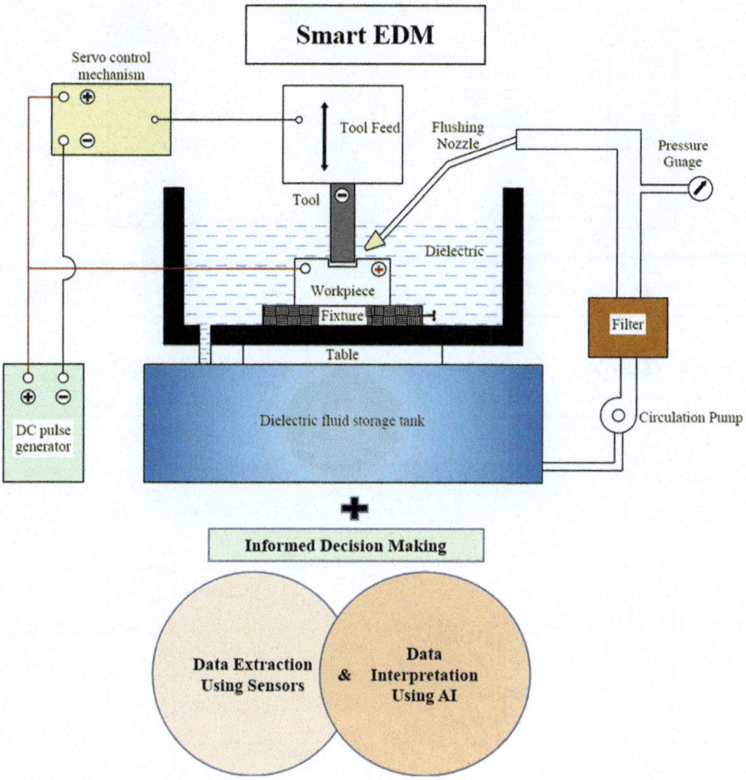

Fig. 4.4 Schematic diagram of the smart EDM

under meticulous control, with parameters such as current, voltage, and the interelectrode gap being precisely managed to achieve the desired machining outcome.

The controlled electrical discharges in the EDM initiate a precise material removal process. As the tool and the workpiece are meticulously brought into close proximity, a potential difference is established by the applying voltage. The carefully managed increase in potential difference prompts a breakdown in the dielectric fluid that envelopes the tool and the workpiece.

This breakdown gives rise to controlled sparks or electrical discharges, symbolizing a burst of energy meticulously orchestrated for material transformation. Figure 4.5a illustrates the initiation of the spark between the tool and the workpiece. The localized high temperature generated by these discharges induces rapid heating, melting, and vaporization of the material, showcasing a moment of spark-induced transformation.

The ensuing drop in voltage across the spark gap paves the way for the passage of current. Figure 4.5b visualizes the localized melting caused by the spark, leading to

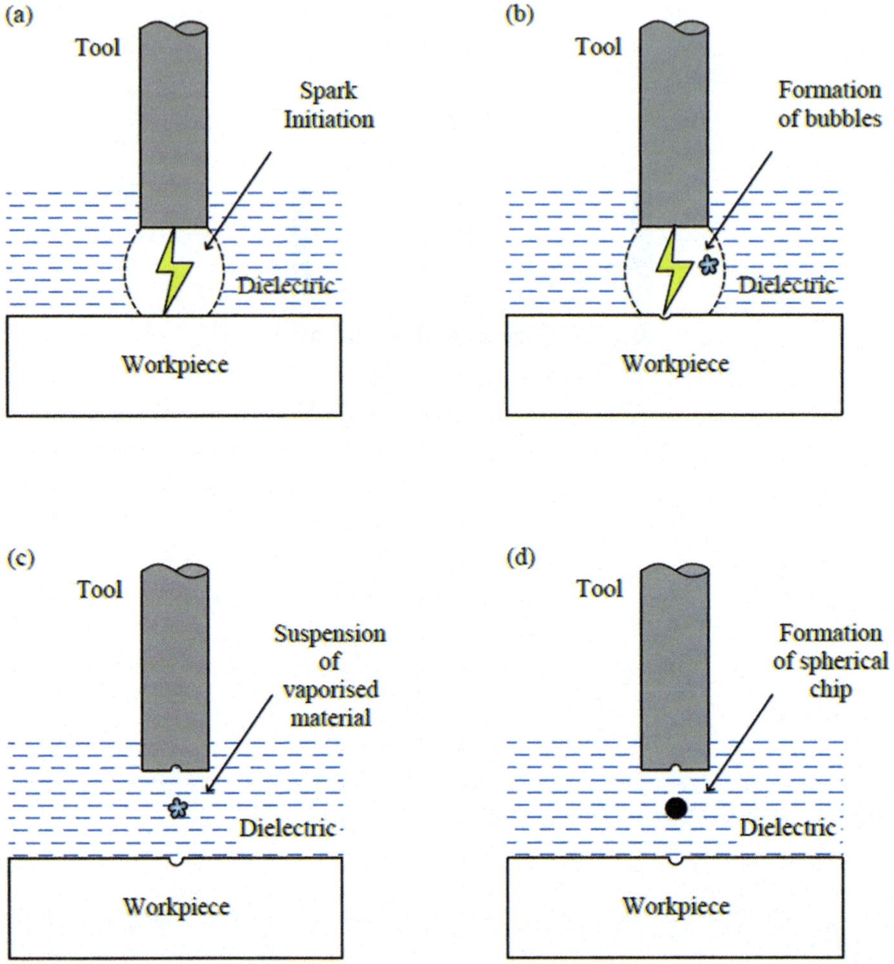

Fig. 4.5 (**a**) Spark initiation between the tool and the workpiece. (**b**) Localized melting by the spark. (**c**) Debris formation. (**d**) Debris flushing by the dielectric fluid

the controlled formation of debris as depicted in Fig. 4.5c. This debris, a by-product of the material removal process, is efficiently flushed away by the dielectric fluid, as demonstrated in Fig. 4.5d.

The small, controlled increments in the material removal play a pivotal role in shaping aerospace alloys with extraordinary precision. The entire EDM process operates under stringent control, with parameters such as voltage, current, and the gap between the tool and the workpiece being meticulously managed. This precision-driven approach ensures that the desired machining outcome is achieved with exacting standards, making EDM a cornerstone in the aerospace engineering for crafting intricate components.

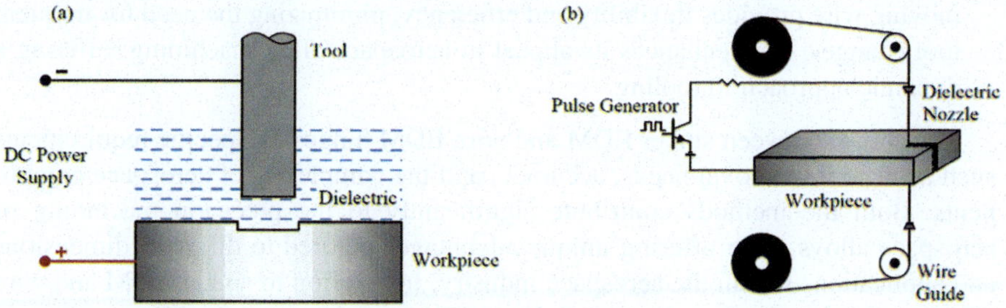

Fig. 4.6 (**a**) Sinker EDM and (**b**) wire EDM

There are two main types of EDM: (i) sinker EDM and (ii) wire EDM. Sinker EDM, also known as die-sinking EDM or conventional EDM, is a precision machining method that utilizes a shaped electrode to erode aerospace alloy workpieces. Often employed for applications demanding intricate shapes and molds in aerospace components, sinker EDM excels in creating three-dimensional geometries with exceptional precision. The shaped electrode, as shown in Fig. 4.6a, typically made of conductive materials such as copper or graphite, imparts its form to aerospace alloys through controlled electrical discharges. This method is especially valuable for crafting intricate shapes and profiles for aerospace applications. Wire EDM, alternatively, relies on a thin wire electrode (50–250 microns) as shown in Fig. 4.6b, to cut through aerospace alloy workpieces with meticulous precision. Guided by computer numerical control (CNC), the continuous movement of the wire allows the creation of intricate shapes and complex profiles in aerospace components. Wire EDM is highly effective in applications where aerospace alloys demand tight tolerances, intricate geometries, and rapid production. Its flexibility and speed make it a preferred choice for various aerospace machining needs. The comparative insights of aerospace alloy machining using sinker EDM and wire EDM are listed below:

- Material Removal Efficiency: Sinker EDM excels in efficiently removing material from aerospace alloys, proving particularly beneficial for crafting molds with intricate details. In contrast, wire EDM demonstrates precision in material removal along with complex contours without the need for frequent tool changes.
- Accuracy and Detailing: When it comes to accuracy and detailing, sinker EDM stands out, offering superior precision and detailing. This makes it an ideal choice for crafting intricate molds and dies for aerospace components. On the other hand, wire EDM, with its precision and speed, is well suited for aerospace alloy parts that demand intricate geometries.
- Application Focus: Sinker EDM finds common application in aerospace alloy machining where depth and intricate molds are crucial for component manufacturing. In comparison, wire EDM has specific utility in aerospace machining scenarios where intricate shapes and tight tolerances are paramount.
- Tooling Dynamics: Sinker EDM may require custom tooling for each job due to its utilization of shaped electrodes. Conversely, wire EDM with its continuously

moving wire provides flexibility and efficiency, minimizing the need for frequent tool changes. This enhances its appeal in aerospace alloy machining, offering a dynamic approach to tooling.

The choice between sinker EDM and wire EDM hinges on specific requirements such as material removal needs, accuracy, and the complexity of aerospace components. Both the methods contribute significantly to the precision machining of aerospace alloys, each offering unique advantages tailored to different dimensions and applications within the aerospace industry. In addition to sinker EDM and wire EDM, it is worth noting that the field of EDM encompasses various specialized variants such as micro-EDM, small-hole EDM. Each variant caters to specific machining requirements, offering unique advantages in precision manufacturing for diverse aerospace applications. While the specific processes may vary, it is noteworthy that the fundamental principle of EDM remains consistent across its different variants. The controlled application of electrical discharges for precise material removal defines the essence of EDM, ensuring a common ground in the pursuit of precision across diverse aerospace machining needs.

4.3 Smart EDM with the Integration of AI

Table 4.1 provides an overview of recent studies, featuring diverse AI techniques such as adaptive neuro-fuzzy inference system (ANFIS), artificial neural networks (ANN), convolutional neural networks (CNN), support vector regression (SVR), and deep neural networks (DNN), employed to optimize critical parameters in EDM. From predicting wire breakages to enhancing surface roughness and depth of cut, these studies showcase innovative approaches and their effectiveness in addressing key challenges within the aerospace alloy machining domain.

EDM is an advanced machining process, and its performance is intricately governed by a series of electrical discharges that illuminate the machining environment for mere fractions of a second. This rapid succession of electrical discharges, each lasting only microseconds, gives rise to a dynamic machining environment characterized by high-frequency events. In the nuanced landscape of EDM, where precision is paramount, the need for sensors becomes indispensable. Sensors serve as vigilant observers, capturing the momentary sparks unfolding in microseconds. These sensors, finely tuned to the rapid and precise changes in electrical potential and current flow, act as crucial instruments for understanding, monitoring, and optimizing the complex dynamics of EDM. They offer real-time insights, ensuring that the machining process unfolds with meticulous precision, contributing to the creation of intricate aerospace components.

Voltage and current sensors stand as the primary instruments, providing comprehensive insights into the machining process. These sensors play a pivotal role because the key elements in the EDM process are intricately linked to electrical current dynamics. By capturing variations in voltage and the flow of current, these

Table 4.1 EDM of aerospace alloys with the integration of AI algorithms and models

References	Material	AI algorithm	Remarks
Abhilash et al. [5]	Inconel 718	ANFIS	ΔV_m predicts wire breakages ANFIS precisely forecasts ΔV_m, ensuring reliable alerts
Abhilash et al. [6]	Inconel 718	ANN	Wire EDM pulse classification model High prediction accuracy using ANN
Yilmaz et al. [7]	Inconel 718 and Ti-6Al-4V	ANFIS + optimization	Development of intelligent system for EDM hole drilling Integrated ANFIS and optimization system
Ozkavak et al. [8]	Inconel 718	Gene expression programming (GEP) and ANN	Predictive model for surface roughness ANN outperforming GEP in prediction accuracy
Ulas et al. [9]	AA7075	Extreme learning machine (ELM) and SVR	Developed ML models to predict surface roughness Best performance weighted—ELM model
Mu et al. [10]	TZM	Intelligent EDM	Improved EDM abilities for extreme machining conditions
Wang et al. [11]	Inconel 718	K-means and hierarchical clustering	Novel tolerance monitoring approach Avoids time-consuming experiments
Nain et al. [12]	Udimet L605	Random forest (RF) and M5P + particle swarm optimization (PSO)	Sensitivity analysis to determine the influential parameter
Wang et al. [13]	Inconel 718	DNN	Detecting tolerance defects in turbine disk fir-tree slots
Ji et al. [14]	Ti-6Al-4V	ANFIS	ANFIS is applied for predicting the depth of cut
Yusoff et al. [15]	Ti48Al	ANN and genetic algorithm (GA)	OrthoANN–media and public relation (MPR)–multiGA approach demonstrates effectiveness

sensors offer a wealth of information, serving as indispensable tools for understanding and optimizing the complex EDM environment. In addition to these sensors, temperature and pressure sensors also provide insights for optimal monitoring of thermal dynamics and the dielectric fluid behavior of the EDM process. Table 4.2 gives a range of sensors employed for data collection along with corresponding descriptions.

The fusion of EDM and AI relies on the synchronized precision facilitated by data acquisition (DAQ). Playing a central role in smart EDM, data acquisition (DAQ) orchestrates the seamless synchronization of information, enabling real-time

Table 4.2 Sensors and their description in EDM

Sensor	Description
Voltage	It measures electrical potential between the tool and the workpiece, providing insights into discharge characteristics
Current	It monitors the flow of electrical current, crucial for understanding material removal and overall process stability
Temperature	It tracks variations in temperature to monitor thermal dynamics and prevent overheating during the EDM process
Pressure	It measures dielectric fluid pressure, offering insights into fluid behavior and ensuring a consistent machining environment
Machine vision camera	It utilizes visual data to inspect workpiece quality, monitor tool wear, and provide a comprehensive view of the machining area.

decisions and predictive insights. The data collected by the sensors are extensive in number. At its essence, DAQ serves as the conductor, gathering and deciphering the data generated by sensors in the EDM environment. The high-frequency events, measured in microseconds, require a sophisticated approach to efficiently capture, process, and utilize information. DAQ systems play a crucial role in bridging the gap between the swift dynamics of EDM and the analytical capabilities of AI. The scope of data acquisition extends beyond individual sensor readings. DAQ systems aggregate data from various sensors, including voltage, current, temperature, pressure, and machine vision cameras. This comprehensive dataset serves as the foundation for AI algorithms to decipher intricate patterns, optimize machining parameters, and anticipate potential deviations in real time. The fusion of data into AI through DAQ holds the potential to overcome the traditional limitations of the EDM process. Real-time adjustments, guided by insights derived from continuous data streams, redefine precision in aerospace alloy manufacturing. Predictive capabilities, fueled by historical data trends and machine learning algorithms, anticipate challenges, enhance efficiency, and contribute to the evolution of aerospace alloy machining.

At the heart of AI integration lies a set of core principles that redefine how aerospace alloy machining unfolds:

(i) Learning algorithms: AI algorithms are the backbone of the integration, continuously learning from the vast datasets acquired through sensors and data acquisition systems. These algorithms adapt and evolve, refining their understanding of intricate machining dynamics over time.

(ii) Real-time decision making: The AI integration enables real-time decision making. As data streams in from sensors and DAQ systems, AI algorithms swiftly analyze, interpret, and make instant adjustments to optimize machining parameters. This dynamic responsiveness enhances precision and efficiency.

(iii) Predictive insights: AI's predictive capabilities enable the foresight. Historical data trends, coupled with machine learning algorithms, empower the system to anticipate challenges before they arise. This proactive approach minimizes disruptions, enhances overall efficiency, and contributes to the continuous evolution of aerospace alloy machining.

(iv) Adaptive optimization: The AI integration fosters an environment of adaptive optimization. Machining parameters, tool path strategies, and other variables are dynamically adjusted based on real-time feedback. This adaptability ensures optimal performance even in the face of changing conditions.

The integration of EDM and AI goes beyond mere combination, unlocking a realm of possibilities for aerospace alloy machining:

- Enhanced precision: AI's analytical prowess, coupled with real-time adjustments, refines precision to unprecedented levels. The system continually adapts its understanding of material removal dynamics, surface finishes, and intricate geometries.
- Efficiency overhaul: Real-time decision making and adaptive optimization lead to a comprehensive overhaul of efficiency in the machining process. Unnecessary delays are minimized, and machining parameters are fine-tuned for the optimal results.
- Cost-reduction: The predictive capabilities of AI contribute to substantial cost-reduction. Anticipating potential issues and optimizing processes prevent material wastage, reduce downtime, and enhance overall resource utilization.
- Versatility in complexity: The integration empowers aerospace engineers to tackle increasingly complex machining tasks. AI's ability to decipher intricate patterns and adapt to diverse scenarios positions it as a valuable ally in crafting components with intricate details and challenging geometries.

Figure 4.7 illustrates the interplay between EDM, sensors, data acquisition, AI algorithms, and outputs in a continuous cycle, encapsulating the essence of smart EDM. This visual representation not only serves as a roadmap for precision in aerospace alloy machining but also highlights the perpetual evolution and refinement integral to the synergy between EDM and AI.

4.4 A Case Study: Estimation of Workpiece Height in Smart Wire EDM

In wire EDM, the estimation of workpiece height or thickness is of paramount importance. To better illustrate the problem, let us consider pulse on time, a key parameter representing the duration of electrical discharges, directly influences the material removal and consecutively the final dimensions of the workpiece. When dealing with workpieces of different thicknesses, pulse on time undergoes intricate variations. In thinner sections, shorter pulse on time values may suffice for efficient material removal, ensuring swift and precise machining. Conversely, thicker workpieces demand longer pulse durations to accommodate the increased volume of material to be eroded.

Aerospace components often require machining of parts having varying thickness. Although modern EDMs are equipped with servo mechanism in order to ensure

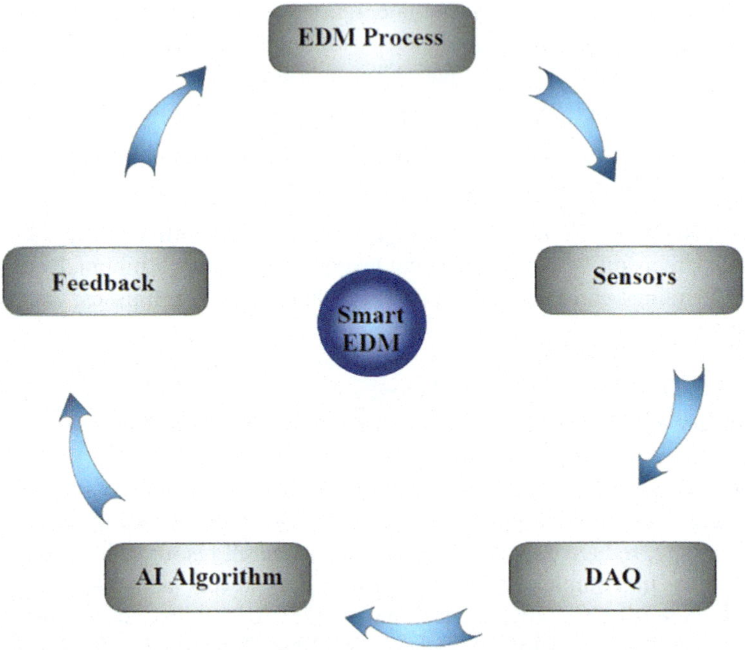

Fig. 4.7 The synergistic cycle of smart EDM

uniform surface morphology throughout the workpiece, adjusting the process parameters according to the height of the workpiece ensures uniform surface finish and prevents process failures like wire breakage.

4.4.1 Experimental Details

Figure 4.8 shows the wire EDM equipped with a high-voltage differential probe for voltage measurements between the wire and the workpiece. The most versatile aerospace alloy Inconel 718 is chosen for the study. Figure 4.9a shows a triangular workpiece having 10 mm × 10 mm × 10 mm dimensions with an inclination of 45°. Figure 4.9b shows the wire electrode that travels from the lower thickness to higher thickness of the workpiece. Since our aim to achieve minimum surface roughness throughout the profile, the process parameters given in Table 4.3 are chosen based on pilot experiments to prevent wire breakages in the process.

A differential probe with a bandwidth of 50 MHz is used to acquire voltage readings between the workpiece and wire electrode while machining. Moreover, a mixed domain oscilloscope with a record length of 10 million points and having a

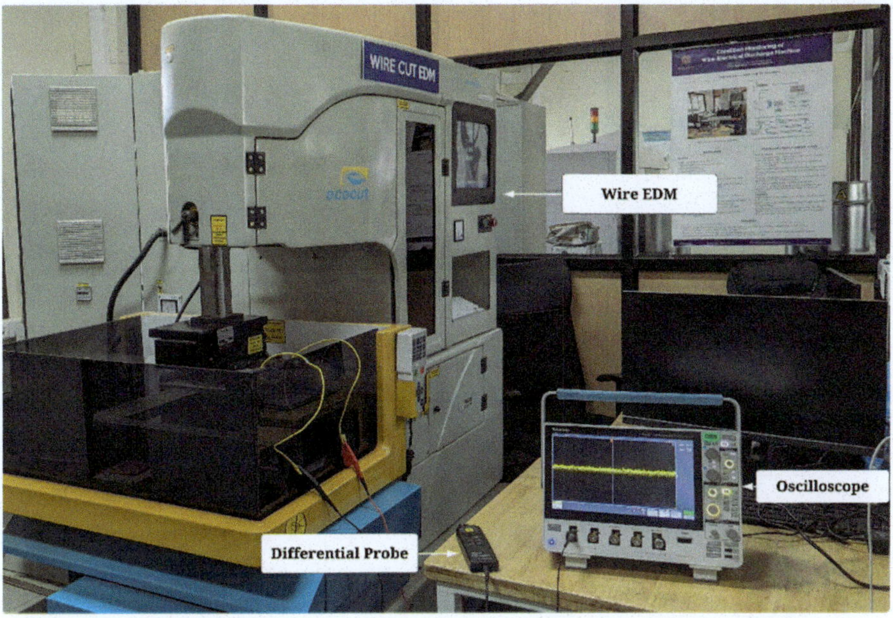

Fig. 4.8 Wire EDM setup equipped with differential probes and oscilloscope

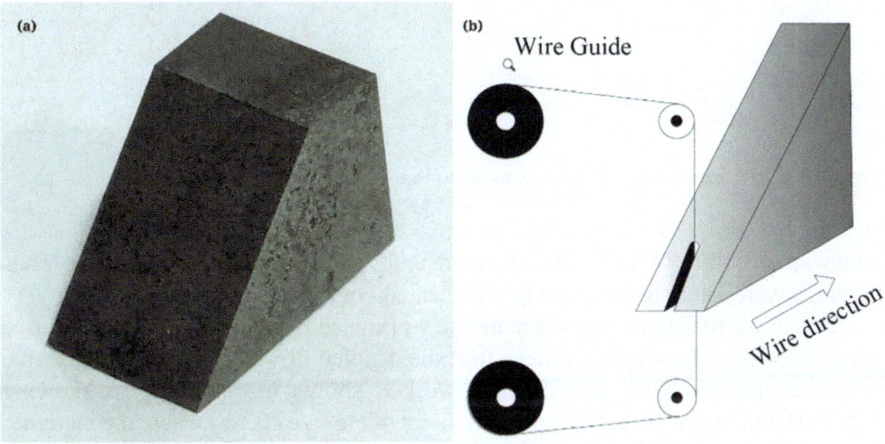

Fig. 4.9 (**a**) Inconel 718 workpiece. (**b**) Schematic illustration of wire direction

Table 4.3 Process parameters in the wire EDM

Pulse-on time	Pulse-off time	Servo voltage	Wire feed
105 μs	50 μs	40 V	5 m/min

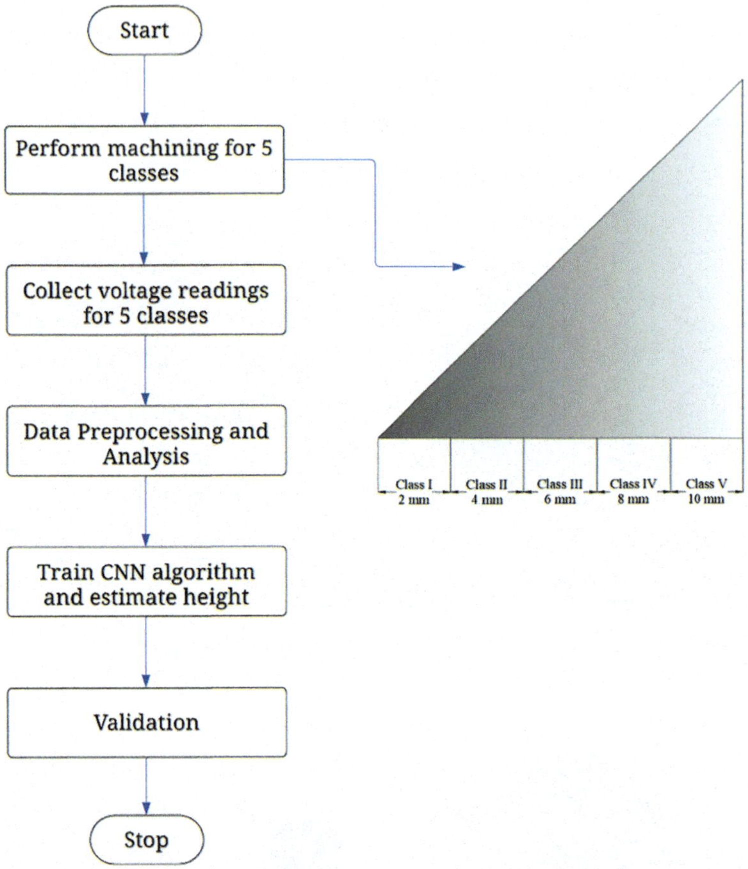

Fig. 4.10 Flowchart in the experiments to determine the workpiece height

bandwidth of 200 MHz is used to read and record the voltage readings. These readings were later analyzed using a dedicated computer.

The aim of the study is to estimate the workpiece height in the wire EDM using AI technologies. In order to achieve this, the detailed flowchart having the activities followed throughout the experimental work is shown in Fig. 4.10. As shown in Fig. 4.10, the height of the workpiece is divided into five classes where the increment of workpiece height is 2 mm. Therefore, the machining is performed for 2 mm, the voltage sensor connected to the wire EDM collects the voltage readings and the oscilloscope is used to read and store the data.

The voltage readings, acquired through the differential probe over a short duration, are depicted in Fig. 4.11. To prepare the data for effective training using convolutional neural networks (CNN), a vital step involves transforming the two-dimensional voltage graphs into meaningful images. This transformation not

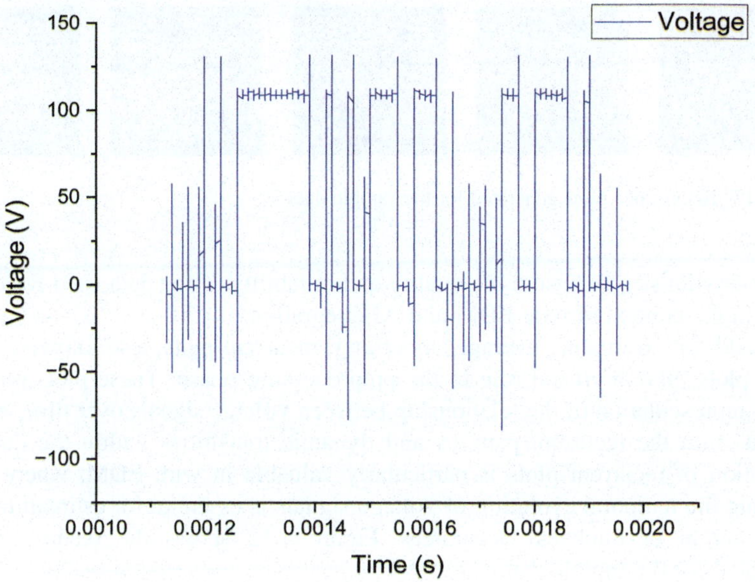

Fig. 4.11 Voltage readings acquired using differential probes

only aids in presenting the data in a visual format but also encapsulates the dynamic transitions in voltage signals corresponding to various machining conditions.

In the data preprocessing phase, moving average smoothing is a critical step to enhance the clarity of voltage signals, aiding in their effective representation for CNN training. Moving average smoothing is employed in the data preprocessing phase to enhance the clarity of voltage signals for several reasons. First, it helps to reduce noise and fluctuations in the data, providing a clearer representation of the underlying trends. This is particularly important in the context of voltage signals obtained during the wire EDM, where the presence of noise can be attributed to various factors such as electrical interference, environmental conditions, and inherent variability in the machining process. Second, moving average smoothing aids in capturing the overall pattern and dynamic transitions in the voltage signals. By averaging the values over a specified window size, it helps to highlight trends and variations that are essential for understanding the impact of different machining conditions on the voltage readings. This is crucial in the context of aerospace alloy machining, where precision in capturing subtle changes in the EDM process is paramount.

In the context of CNN training, the use of smoothed voltage signals as input data enhances the network's ability to learn and generalize patterns. The reduced noise and enhanced clarity contribute to a more effective training process, ultimately leading to improved accuracy in estimating workpiece height or thickness based on the voltage readings. Overall, moving average smoothing serves as a valuable

Fig. 4.12 Recurrence plots generated for the sample data

preprocessing step to ensure the quality and reliability of the data used for training CNN in the context of wire EDM and AI integration.

In addition to moving average smoothing, recurrent plots, also known as recurrence plots, play a crucial role in the preprocessing phase. These plots provide a visual representation of the relationship between voltage signals over time, offering insights into the recurrent patterns and dynamic transitions within the data. The utilization of recurrent plots is particularly valuable in wire EDM, where understanding the temporal evolution of voltage signals is essential for estimating workpiece height or thickness accurately. Figure 4.12 shows the recurrence plots generated for the sample data.

The recurrent plots are generated by calculating pairwise distances between data points in chunks, capturing the similarity or dissimilarity between voltage signals at different time points. The resulting plots highlight regions where the voltage signals exhibit recurrent patterns, aiding in the identification of underlying trends and periodicities. This is pivotal for aerospace alloy machining, where the stability and consistency of the EDM process are critical for achieving precision in workpiece dimensions.

By visualizing the recurrence patterns, recurrent plots contribute to the interpretability of the voltage signals and provide a foundation for CNN training. The patterns captured in these plots serve as informative features for the network, enhancing its ability to recognize and learn the relationships between voltage signals and workpiece height.

In the estimation of workpiece height in wire EDM, we employ a CNN enriched by the insights gained from earlier phases of data preprocessing. Our experiment encompasses five distinct classes, each representing a specific increment of 2 mm in workpiece height. A meticulously curated training set is created for the five classes to ensure diverse representation and adequate learning.

One noteworthy aspect of our approach is the utilization of transfer learning, a technique where knowledge gained from a preexisting model is adapted to our specific task as shown in Fig. 4.13. We leverage the power of ResNet 50, a pretrained deep learning model known for its prowess in image-related tasks. The pretraining on large-scale image datasets enables ResNet 50 to learn intricate features and patterns that are transferrable to our voltage signal images, simplifying thc training process and enhancing the model's ability to generalize.

The CNN's architecture facilitates the automatic extraction of hierarchical features from the transformed voltage signal images. Through the training process, the model learns to associate these features with the five classes of workpiece height,

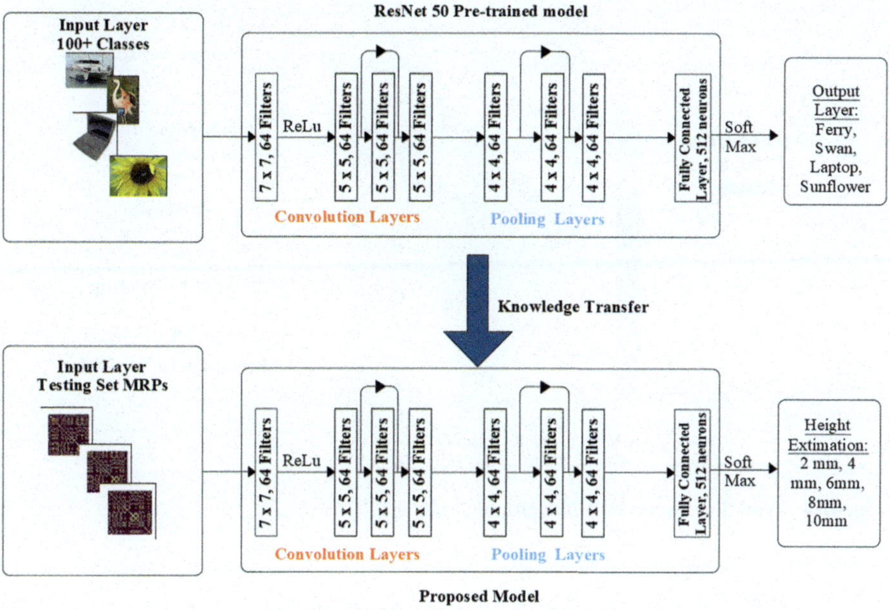

Fig. 4.13 CNN architecture for height estimation in the wire EDM

effectively discerning the nuances of the machining conditions. The use of transfer learning allows our CNN to adapt its knowledge to the specific challenges posed by wire EDM, making it a potent tool for workpiece height estimation.

Upon training completion, the CNN's performance is rigorously evaluated on a dedicated test set, ensuring its capability to generalize to new, unseen data. The trained model serves as a powerful tool for predicting workpiece height across various machining scenarios.

4.4.2 *Results and Discussion*

Figure 4.14 shows the workflow overview of the proposed model. As shown in the schematic, after the training of the five classes based on the workpiece height, a similar workpiece is machined and the obtained machining recurrent plots (MRPs) are tested using the proposed CNN model and the workpiece height is estimated.

To thoroughly evaluate the model's performance, we utilize two essential metrics: (i) confusion matrix and (ii) receiver operating characteristic (ROC) curve.

The confusion matrix illustrates the classification performance of the model. In this case, an overall prediction accuracy of 86% is observed, with 86 instances correctly classified out of 100 across different classes. Figure 4.15 depicts the confusion matrix.

Table 4.4 summarizes the performance metrics for each class. The results reveal a strong overall accuracy in predicting the height of each class, underscoring the

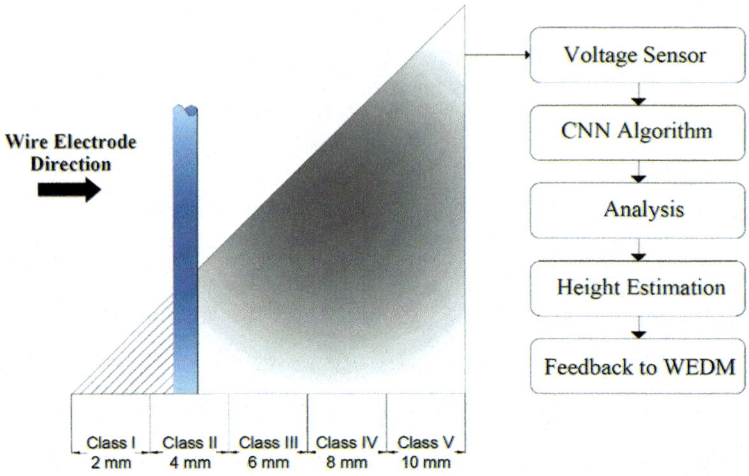

Fig. 4.14 Workflow overview of the proposed model

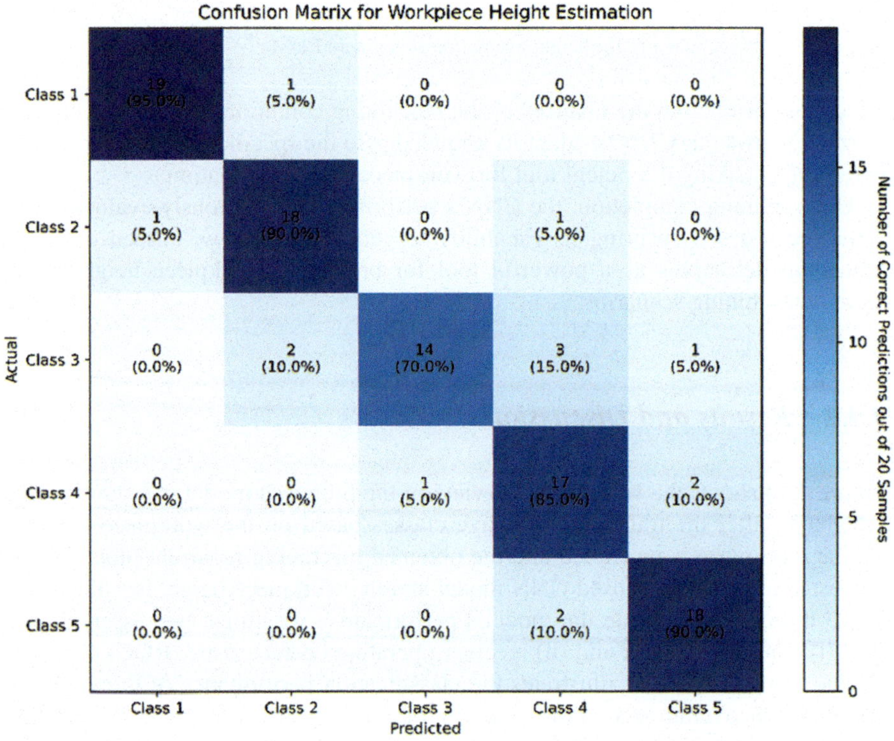

Fig. 4.15 Confusion matrix for the proposed model

Table 4.4 Performance metrics for each class

Class	Accuracy	Precision	Recall	F1-score
I	0.98	0.95	0.95	0.95
II	0.95	0.85	0.90	0.87
III	0.93	0.93	0.70	0.80
IV	0.91	0.73	0.85	0.79
V	0.95	0.85	0.90	0.87

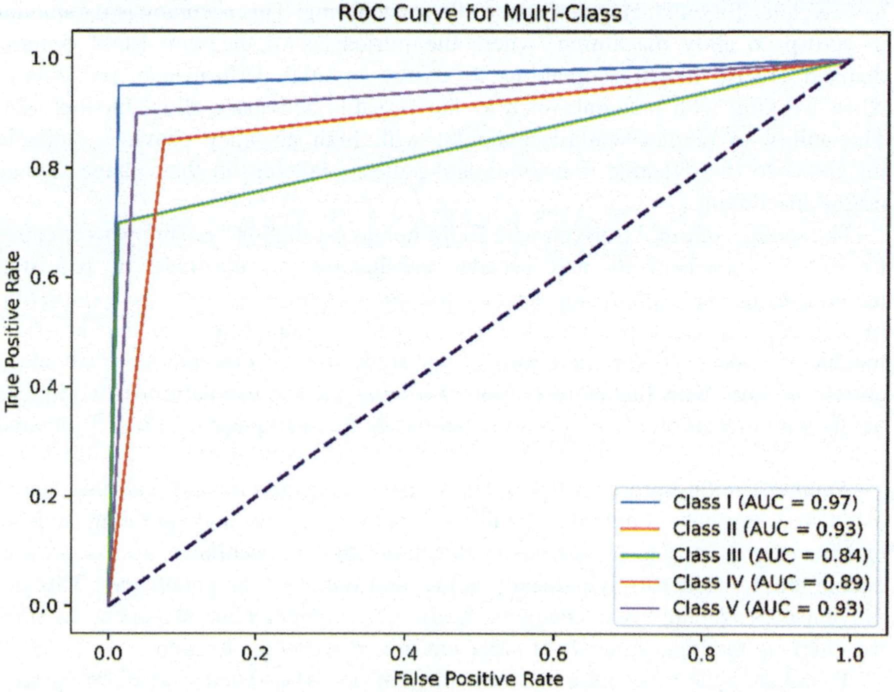

Fig. 4.16 ROC curve for the proposed model

efficacy of the model in workpiece height estimation. However, it is noteworthy that the current model encounters challenges in accurately predicting workpiece height below 2 mm, particularly evident in Class III. This limitation is attributed to specific sensor characteristics. Future works might involve integrating data from a different sensor, offering more information and the potential to improve height estimation for dimensions beyond 1 mm.

The ROC curve, depicted in Fig. 4.16, serves as an additional validation of the model's performance. Notably, the area under the curve (AUC) for each class demonstrates strong discriminatory power: Class I (AUC = 0.97), Class II (AUC = 0.93), Class III (AUC = 0.84), Class IV (AUC = 0.89), and Class V (AUC = 0.93). These AUC values highlight the ability of the model to distinguish between classes, with high true positive rates and relatively low false positives. While these results indicate a commendable level of accuracy, it is important to note that the current model encounters challenges in predicting workpiece heights below 2 mm, attributed to specific sensor characteristics. Future endeavors involve addressing this limitation by integrating data from a current sensor, aiming for enhanced height estimation capabilities, particularly for dimensions under 1 mm.

The successful integration of AI technologies into the wire EDM process not only demonstrates remarkable accuracy in predicting workpiece height but also opens new avenues for enhancing aerospace alloy machining. This accuracy is paramount in aerospace alloy machining, where the intricacies of the wire EDM process demand precise height estimations to ensure optimal performance, particularly when working with materials such as the versatile aerospace alloy, Inconel 718. The ability to predict workpiece height with high accuracy provides valuable insights into the dynamic transitions and patterns inherent in the voltage signals during machining.

The success of our AI-driven wire EDM height estimation system opens avenues for future enhancements and broader applications in the field of precision manufacturing. One promising avenue for improvement involves the exploration of alternative sensors, particularly those capable of capturing finer details in the machining process. Future iterations of our study may incorporate more advanced current sensors with higher resolution, allowing for the estimation of workpiece height with greater precision, even for estimating the workpiece height of 1 mm and below.

Another exciting prospect lies in advancing our system toward real-time height estimation through continuous condition monitoring. By incorporating sensors capable of providing instantaneous feedback during the machining process, we can enable the AI model to dynamically adjust and optimize the parameters. This not only ensures adaptability to changing machining conditions but also opens the door to achieving the highest levels of precision in real-world applications.

To enhance the robustness and versatility of our AI-assisted wire EDM system, we may explore the integration of multiple sensors, each capturing distinct aspects of the machining environment. Fusion of data from different sensors, such as voltage sensors and current sensors, could provide a comprehensive understanding of the machining conditions, further improving the accuracy and reliability of workpiece height estimations.

Efforts can be directed toward scaling our AI-driven system for widespread industry adoption. Collaboration with manufacturing entities and incorporating feedback from real-world applications are crucial for refining the model's generalizability across diverse manufacturing environments. In essence, the future holds exciting possibilities for refining our AI-assisted wire EDM height estimation system. By pushing the boundaries of sensor technology, embracing real-time monitoring, and fostering collaboration with industry stakeholders, we aim to create a platform that not only sets new standards in precision manufacturing but also adapts seamlessly to the evolving needs of the aerospace industry.

4.5 Conclusions

In summary, the integration of AI into EDM in aerospace manufacturing represents a profound stride in precision engineering. The processes with a range of vigilant sensors, including voltage, current, temperature, pressure sensors, and machine vision cameras, establish a robust foundation for meticulous manufacturing processes. The orchestration of data acquisition acts as a conductor, harmonizing the velocity of EDM with the analytical skill of AI, thereby reshaping the aerospace alloy machining landscape. The successful application of diverse AI techniques, particularly in wire EDM workpiece height estimation using CNN, showcases the transformative potential of this integration.

As we look ahead, the promise of advanced sensors, real-time estimation, and the integration of multisensor fusion hints at a future where precision and adaptability in aerospace manufacturing reach new heights. This union of EDM and AI not only transcends mere technological convergence but also charts a course for the future of aerospace engineering, where every spark in the EDM process resonates with the sophistication of AI algorithms.

Acknowledgments We sincerely thank the Department of Mechanical Engineering, IIT Palakkad, for their invaluable support and resources, essential to the completion of this book chapter.

References

1. Abhilash PM, Chakradhar D. Surface integrity comparison of wire electric discharge machined Inconel 718 surfaces at different machining stabilities. Procedia CIRP. 2020;87:228–33. https://doi.org/10.1016/J.PROCIR.2020.02.037.
2. Ho KH, Newman ST. State of the art electrical discharge machining (EDM). Int J Mach Tools Manuf. 2003;43(13):1287–300. https://doi.org/10.1016/S0890-6955(03)00162-7.
3. Kliuev M, et al. EDM drilling and shaping of cooling holes in Inconel 718 turbine blades. Procedia CIRP. 2016;42:322–7. https://doi.org/10.1016/J.PROCIR.2016.02.293.
4. Abhilash PM, Chakradhar D. Prediction and analysis of process failures by ANN classification during wire-EDM of Inconel 718. Adv Manuf. 2020;8:519–36. https://doi.org/10.1007/s40436-020-00327-w.
5. Abhilash PM, Chakradhar D. ANFIS modelling of mean gap voltage variation to predict wire breakages during wire EDM of Inconel 718. CIRP J Manuf Sci Technol. 2020;31:153–64. https://doi.org/10.1016/J.CIRPJ.2020.10.007.
6. Abhilash PM, Chakradhar D, Luo X. Machine learning based classification and analysis of wire-EDM discharge pulses. In: 2022 8th international conference on nanomanufacturing & 4th AET symposium on ACSM and digital manufacturing (Nanoman-AETS). IEEE; 2022. pp. 1–6, doi: https://doi.org/10.1109/NANOMAN-AETS56035.2022.10119466.
7. Yilmaz O, Bozdana AT, Okka MA. An intelligent and automated system for electrical discharge drilling of aerospace alloys: Inconel 718 and Ti-6Al-4V. Int J Adv Manuf Technol. 2014;74(9–12):1323–36. https://doi.org/10.1007/S00170-014-6059-1/METRICS.
8. Varol Ozkavak H, Sofu MM, Duman B, Bacak S. Estimating surface roughness for different EDM processing parameters on Inconel 718 using GEP and ANN. CIRP J Manuf Sci Technol. 2021;33:306–14. https://doi.org/10.1016/J.CIRPJ.2021.04.007.

9. Ulas M, Aydur O, Gurgenc T, Ozel C. Surface roughness prediction of machined aluminum alloy with wire electrical discharge machining by different machine learning algorithms. J Mater Res Technol. 2020;9(6):12512–24. https://doi.org/10.1016/J.JMRT.2020.08.098.
10. Mu X, Zhou M, Zhang J, Lu N, Ye Q. Intelligent electrical discharge machining (EDM) molybdenum titanium zirconium alloy by an extended adaptive control system. J Manuf Process. 2022;77:207–18. https://doi.org/10.1016/J.JMAPRO.2022.03.003.
11. Wang J, Sanchez JA, Ayesta I, Iturrioz JA. Unsupervised machine learning for advanced tolerance monitoring of wire electrical discharge machining of disc turbine fir-tree slots. Sensors. 2018;18(10):3359. https://doi.org/10.3390/S18103359.
12. Nain SS, Garg D, Kumar S. Performance evaluation of the WEDM process of aeronautics super alloy. Mater Manuf Process. 2018;33(16):1793–808. https://doi.org/10.1080/10426914.2018.1476761.
13. Wang J, Sanchez JA, Iturrioz JA, Ayesta I. Geometrical defect detection in the wire electrical discharge machining of fir-tree slots using deep learning techniques. Appl Sci. 2019;9(1):90. https://doi.org/10.3390/APP9010090.
14. Ji M, Muthuramalingam T, Saravanakumar D, Karmiris-Obratański P, Karkalos NE, Zhang W. Predicting depth of cut in vibration-assisted EDM cutting on titanium alloy using adaptive neuro fuzzy inference system. Measurement. 2023;219:113245. https://doi.org/10.1016/J.MEASUREMENT.2023.113245.
15. Yusoff Y, Zain AM, Amrin A, Sharif S, Haron H, Sallehuddin R. Orthogonal based ANN and multiGA for optimization on WEDM of Ti–48Al intermetallic alloys. Artif Intell Rev. 2019;52(1):671–706. https://doi.org/10.1007/S10462-017-9602-2/FIGURES/10.

Chapter 5
Nanoparticle Reinforcement in Friction Stir Welding of Dissimilar Aerospace Alloys

Amin Abdollahzadeh and Behrouz Bagheri Vanani

5.1 Introduction

In aerospace, shipbuilding, and automotive industries, magnesium alloys have become noticeably crucial lightweight structural materials due to their high specific strength and low density [1–3]. Magnesium alloys offer improved fatigue life due to their increased energy absorption behavior, dimensional stability, and excellent castability [4]. On the other hand, aluminum alloys have low specific weight, excellent thermal and electrical conductivity, high corrosion resistance, and good formability [5]. In addition, the possibility of effectively joining magnesium alloy components to other industrial materials, especially aluminum alloys, increases design flexibility and their applications. Therefore, the improvement of reliable joints between magnesium and aluminum alloys is required to obtain the advantages of joining these two materials [6]. However, traditional fusion joining of magnesium and aluminum alloys leads to the formation of brittle intermetallic compounds (IMCs) and coarse grains in the weld zone [7]. Friction stir welding (FSW) presents a feasible approach to joining dissimilar materials, such as magnesium to aluminum alloys, due to the lack of formation of a melt pool in the weld zone during the solid-state process [8]. The welding parameters to provide sound dissimilar alloy welds with refined microstructure and high mechanical characteristics in magnesium and aluminum alloys have been thoroughly examined. Zhao et al. [9] investigated the

A. Abdollahzadeh
School of Metallurgy and Materials Engineering, Iran University of Science and Technology (IUST), Tehran, Iran
e-mail: am_abdollahzadeh@iust.ac.ir

B. B. Vanani (✉)
Department of Materials and Metallurgical Engineering, Amirkabir University of Technology, Tehran, Iran
e-mail: b.bagheri@aut.ac.ir

S. Gürgen (ed.), *Advances in Manufacturing for Aerospace Alloys*, Sustainable Aviation, https://doi.org/10.1007/978-3-031-64455-9_5

role of welding parameters on dissimilar magnesium and aluminum alloys during the FSW process. They concluded that high joining efficiency and a high shear force were obtained for the sample welded at a joining speed of 70 mm/min. Ding et al. [10] investigated the characteristics of the magnesium solid solution layer of dissimilar magnesium and aluminum joints fabricated by FSW. It was noticed that the transition zone would be divided into five distinct layers, and a coherent interface was fabricated between the magnesium solid solution and magnesium silicide (Mg_2Si) phases.

The addition of other materials, due to inevitable fabrication of hard and brittle IMCs, has been known as a curtail-assisted process during the dissimilar joining [11–13]. On the other hand, the addition of ceramic particles as reinforcing particles within the weld zone has been an acceptable and effective technique to improve the strength of weld samples. Fallahi et al. [14] applied silicon carbide (SiC) reinforcing particles during the FSW of AA5083 and stainless steel. No remarkable formation of defects was observed at the stir zone (SZ), and the joint strength ratio to ultimate tensile strength was improved by around 94%. The influence of ceramic powder on microstructure and mechanical performance of weld specimens during the joining of aluminum and magnesium alloys by FSW was estimated by Abdollahzadeh et al. [7]. It was reported that the hardness of the weld sample, due to more microstructure refinement by the addition of SiC particles, is high, and the fracture surface shows ductile mode deformations. On the other hand, the sample welded free of reinforcing particles has a fracture surface indicating brittle type of crack propagation. The role of reinforcing particle was investigated by Moradi et al. [15] to improve the mechanical and microstructural characteristics of dissimilar alloys, AA2024 and AA6061, during the FSW process. They concluded that the grain size of the weld zone significantly reduces while the mechanical behaviors, namely strength and toughness, increase. Tabasi et al. [16] incorporated the SiC nanoparticles for the dissimilar joining of AZ31 to AA7075 during the FSW operation. They observed that the microstructure evolution significantly occurred in the weld zone, and a metal matrix composite with a uniform distribution of reinforcing particles was achieved. The role of titanium dioxide (TiO_2) reinforcing particles on the microstructural and mechanical characteristics of dissimilar magnesium and aluminum alloys during the FSW process was analyzed by Asl et al. [17]. It was concluded that the weld sample, achieved at 40 mm/min and 800 rpm, presented the highest tensile strength of 132 MPa, which is around two times higher compared with the weld sample fabricated without particle reinforcement. The presence of fine microstructure, thin IMC layers, and uniform distribution of reinforcing particles are other advantages of reinforcing particles during the FSW of dissimilar joints.

It can be concluded that the role of reinforcing nanoparticles on microstructural properties, IMC formations, material flow characteristics, and mechanical properties of aluminum and magnesium dissimilar joints were not investigated properly. The objective of this study is to analyze the feasibility of FSW process for joining AZ91 and AA6061 alloys with the addition of titanium carbide (TiC) nanoparticle reinforcement.

5.2 Materials and Method

In the present study, specimens with 60 mm width and 100 mm length are cut from 5-mm-thick plates of AZ91 and AA6061T6 alloys. Table 5.1 gives the chemical compositions of these alloys. The joining procedure consists of both friction stir process (FSP) and FSW. Regarding FSP, a groove configuration serves as a container for adding nanoparticles. Considering the dimensions of the groove, the designed geometry of the pin, and the 5 mm thickness of the plates, a profile with dimensions of $2 \times 2 \times 90$ mm^3 for width, depth, and length, respectively, was selected on the faying surface of the magnesium alloy. This choice aimed to achieve an equal volume fraction of TiC particles, each with an average size of 60 nm. To effectively secure the nanoparticles between the workpieces, to prevent powder dissipation during FSW, and to ensure an acceptable nanoparticle distribution on the adjoining side of the magnesium sheet, three passes of the FSP were conducted using a pinless tool.

FSW was executed using a machine (FP4ME-CNC) equipped with a spindle power and a maximum speed of 4 kW and 2800 rpm, respectively. The FSW tool, fabricated from H13 steel with a hardness of 50 HRc, was utilized for the welding operation. Table 5.2 provides additional tool details. It is noteworthy that, for the FSP, a pinless tool was employed to close the groove, whereas the frustum pin geometry was employed for the FSW operation.

After a series of trial and error, the optimal microstructure evolution and mechanical characteristics were attained for specimens joined at rotational and traverse speeds of 850 rpm and 45 mm/min, respectively. To investigate the effects of TiC nanoparticles on the microstructural and mechanical characteristics of the joined samples, the FSW operation was also conducted using the same process parameters, excluding the consideration of reinforcing particles. Figure 5.1 shows a schematic view of the FSP and FSW operations employed.

Table 5.1 Chemical compositions of the alloys

	Chemical composition (wt%)								
Material	Al	Mg	Si	Zn	Fe	Cu	Mn	Cr	Ti
AA6061T6	Bal.	0.9	0.64	0.003	0.3	0.25	0.01	0.1	0.02
AZ91	9.10	Bal.	0.09	0.70	0.003	0.0097	0.21	–	–

Table 5.2 Tool details for FSP and FSW

Process	Tool material	Shoulder diameter (mm)	Pin shape	Shoulder surface	Outer dimeter of pin (mm)	Inner diameter of pin (mm)	Pin length (mm)
FSP	H13 steel	15	Pinless	Concave	–	–	–
FSW	H13 steel	15	Frustum	Concave	6	4	4.8

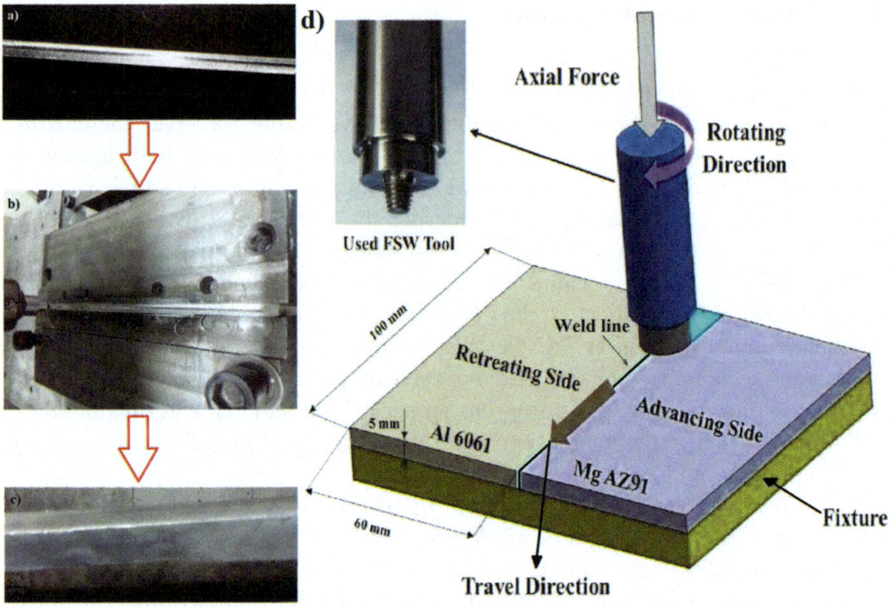

Fig. 5.1 (**a**) Groove configuration for TiC particles, (**b**) FSP with pinless tool, (**c**) sample after FSP, (**d**) schematic illustration of FSP and FSW

For metallographic assessments, cross-sectional samples perpendicular to the FSW direction were prepared, following ASTM-E3 standards for microstructural preparation. Welded samples underwent mechanical polishing up to 4000 grit, followed by polishing with aluminum oxide (Al_2O_3) solution with a particle size of up to 1 μm. For the microstructure analysis of the AZ91 side, weld specimens were etched using a solution consisting of 1 mL acetic acid, 1 mL nitric acid, and 1 mL oxalic acid in 150 mL distilled water for 5 s. Similarly, for the microstructure analysis of the AA6061 side, weld specimens were etched with a solution comprising 3 mL hydrochloric acid, 2 mL hydrofluoric acid, and 5 mL nitric acid in 190 mL distilled water for 60 s. Microstructural evolution was examined using optical microscopy, field emission scanning electron microscopy (FESEM), and high-resolution transmission electron microscopy (HRTEM). For HRTEM analysis, samples were initially cut and ground to a thickness of less than 100 μm using 1000 grit abrasive paper. Subsequently, a 3-mm disc was punched out from the specimen based on the optical micrograph, selecting the region containing IMCs near the advancing side and the interface of magnesium and aluminum. The punched specimens were then polished using diamond paste, followed by ion milling for thinning purposes.

Tensile specimens were prepared in a sub-sized scale, following ASTM-E8 standards, oriented vertically to the FSW direction. Tensile testing was conducted at ambient temperature with a strain rate of 0.5 mm/min using an Instron-5502 testing machine. Microhardness tests were performed across the center of metallographic specimens using a Buehler Micromet-2 system with a 25-g load and 10-s holding time. Each indentation point in this measurement was approximately 0.5 mm apart from adjacent points.

5.3 Results and Discussion

5.3.1 Metallurgical Observations

5.3.1.1 Weld Surface Appearance

Two primary parameters influencing joint strength during the FSW process are macrofeatures, such as tunnel types and cavity defects, and microfeatures, including the generation of IMCs leading to cracking along the interface and fracture surface [18]. Figure 5.2 shows the macrostructure of weld specimens with and without reinforcing particles. It is evident that, without reinforcing particles, the heat input was insufficient to generate adequate material flow during conventional FSW (Fig. 5.2a). Consequently, numerous defects such as microvoids, tunnels, microcracks, and cavities are observed in the weld nugget. Previous studies indicate that a simple contact interface is formed in the upper side of the SZ, while an intermixing microstructure is observed in the lower side of the SZ due to the plastic flow of material within the process [19]. Therefore, the insufficient time for material mobility from the front of the pin tool to the rear increases the likelihood of defect formation such as voids and tunnels. In contrast, as shown in Fig. 5.2b, joint quality improves significantly with sufficient heat input, resulting from internal interaction between reinforcing particles. This interaction enhances frictional heat production, leading to a plasticized SZ and, consequently, increased material flow. Under these conditions, the weld surface morphology is smooth, and optimized heat production and material flow eliminate the formation of defects such as cavities and interface cracks.

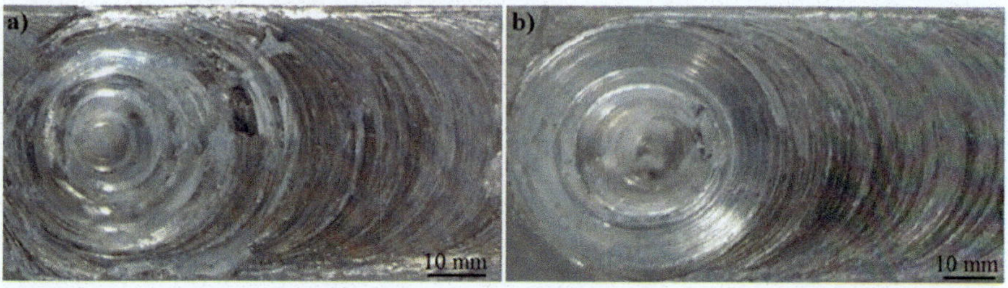

Fig. 5.2 Surface appearance of FSW samples: (**a**) without and (**b**) with reinforcing particles

5.3.1.2 Microstructure Evolution

Figure 5.3 shows the cross-sectional macrostructure of the joint specimen fabricated with reinforcing particles, showing proper metallurgical bonding and a defect-free joint. Following the FSW process, various regions are formed in the weld joint, namely the SZ, thermomechanically affected zone (TMAZ), and heat-affected zone (HAZ) [20]. Due to different plastic flows during FSW, the boundary between SZ and TMAZ differs on the magnesium and aluminum sides. Specifically, material is squeezed from the advancing side to the retreating side and forms the SZ. In our study, there is good alignment between the plastic deformation direction of the weld and the base material, with a relative discrepancy observed between the border of the deformed zone and the base material, making the border between TMAZ and SZ noticeable.

On the retreating side, the deformed material flows behind the pin tool in the opposite direction to the joining direction. In this scenario, material flow occurs simultaneously in the joint center and base material, resulting in an indistinct border between TMAZ and SZ. At the top of the SZ, there is plain intermixing of the two materials, while complex intercalated interlayers resembling onion rings are observed at the bottom of the SZ. The presence of a complex interface of magnesium and aluminum alloys at the SZ enhances both metallurgical bonding and mechanical interlocking, thereby improving the tensile properties of the joint area [21–23]. This underscores the significance of mechanical interlocking in achieving superior performance in welding dissimilar materials, as extensively reported by Huang et al. [24]. Given the higher heat input in the advancing side compared with the retreating side, it appears that intermixing on the magnesium side of the SZ is more pronounced than on the aluminum side, resulting in more intense plastic deformation on the magnesium side.

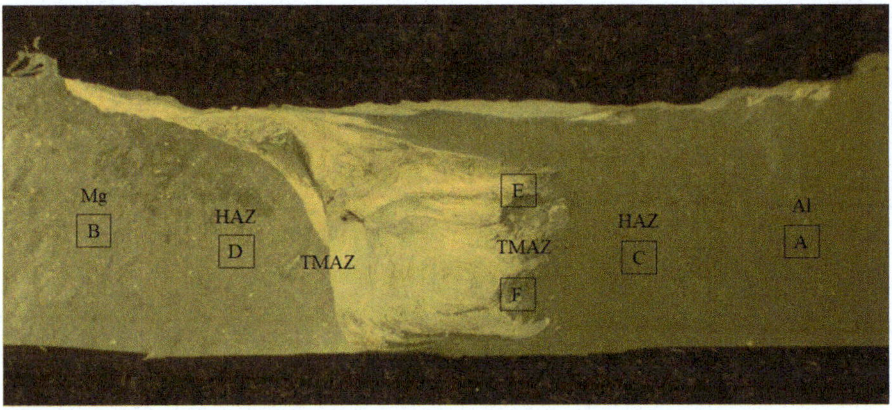

Fig. 5.3 Cross-sectional macrostructure image of the sample with TiC particles

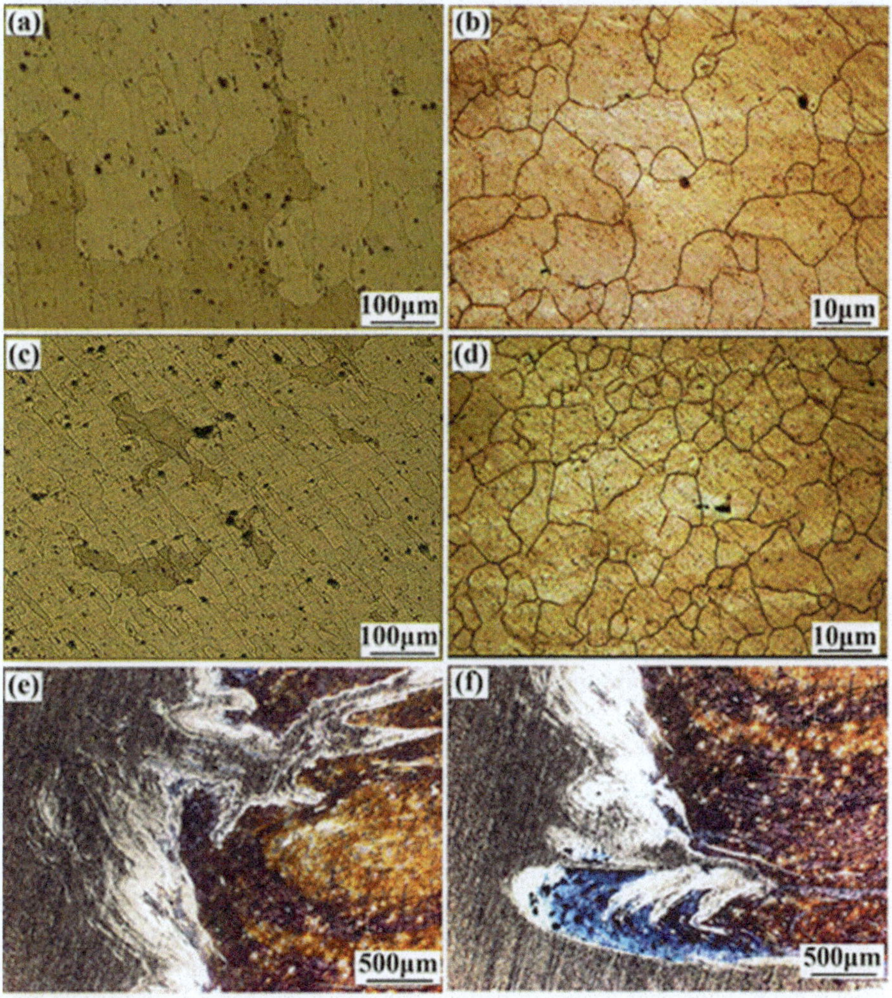

Fig. 5.4 Microstructure of different regions in dissimilar joint with reinforcing particles: (**a**) AA6061T6 base metal, (**b**) AZ91 base metal, (**c**) HAZ in AA6061T6, (**d**) HAZ in AZ91, (**e**) upper side of SZ, and (**f**) lower side of SZ

The microstructure of the AA6061T6 base material, characterized by elongated grains oriented to the rolling direction, and the annealed AZ91 base material, exhibiting coarse equiaxed grains, are depicted in Fig. 5.4a, b, respectively. The HAZ, adjacent to the weld zone, experiences thermal influence during the welding process. As shown in Fig. 5.4c, d, the HAZ grains on both the aluminum and magnesium sides are slightly coarser compared with those of the base materials due to the higher temperatures experienced in the HAZ. During the FSW of dissimilar alloys, intercalated microstructures and lamellar-like bands containing recrystallized grains of aluminum and magnesium alloys are formed [25]. The severe plastic deformation at elevated temperatures induces dynamic recrystallization in the SZ, resulting in the formation of finer recrystallized grains compared with the initial materials. However, dynamic recrystallization does not occur in the HAZ, leading to coarser grains in this region.

The incorporation of reinforcement nanoparticles restricts the mobility of grain boundaries, acting as obstacles against grain growth, and promotes the nucleation of new grains, known as the pinning effect. Consequently, the addition of TiC reinforcements significantly reduces the grain size compared with joint samples without reinforcement, thereby enhancing weld joint strength and toughness [26].

The microstructure of the weld sample in the upper and lower sides of the SZ (Fig. 5.4e, f, respectively) exhibits onion rings, intercalated layers, and a chaotic distribution of fine intermetallic phases. Recrystallized aluminum and magnesium alloys are blended together, forming complex intercalated flow patterns. Energy-dispersive spectroscopy (EDS) analysis by Azizieh et al. [27] reveals that white and grey bands predominantly consist of aluminum and magnesium, respectively. Consequently, dark and light zones likely indicate magnesium-rich and aluminum-rich areas, respectively. The formation of the $Al_{12}Mg_{17}$ phase, more susceptible to corrosion from the etchant solution due to its higher magnesium content, results in darker areas [28]. Conversely, the Al_3Mg_2 phase, with higher aluminum content, appears lighter after etchant corrosion. These compounds form thin, consistent layers, creating a banding structure around the aluminum-rich area.

Figure 5.5 shows FESEM macrostructural and microstructural images of the joint sample fabricated with reinforcing particles, along with EDS analysis of regions denoted as A and B, associated with the TiC-poor and TiC-rich zones, respectively. Chemical composition estimates of the marked areas are provided in Table 5.3, and EDS analysis of the TiC-rich zone (Fig. 5.5d) reveals a prominent titanium peak, indicating the presence of significant amounts of TiC particles in this area. It is noteworthy that agglomeration of reinforcing TiC particles occurred, with these particles exhibiting a tendency to migrate toward the advancing side, indicating a material flow pattern in FSW [29]. However, several studies have suggested that greater grain refinement occurs in particle-rich zones compared with particle-poor zones. Moreover, a higher distribution of nano-sized reinforcements in the SZ contributes to increased pinning of grain boundaries, resulting in finer grains and enhanced tensile strength [30–32].

Figure 5.6 provides a higher magnification view of the TiC-rich area identified as Region B in Fig. 5.5b. The grain size in this area ranges from 50 to 150 nm, although particle agglomeration is also evident due to the stirring action of the FSW tool in the SZ. It is important to note that while a higher homogeneous distribution of nanoparticles contributes to increased mechanical characteristics of the joint specimens, a certain degree of particle agglomeration is unavoidable during the FSW of dissimilar materials [33].

Figure 5.7 shows the FESEM micrograph of the SZ and distribution maps of different elements for weld samples fabricated with reinforcing particles. In Fig. 5.7b, c, aluminum and magnesium elements exhibit more pronounced accumulation patterns, forming a lamellar-like structure with consistent thin layers. In

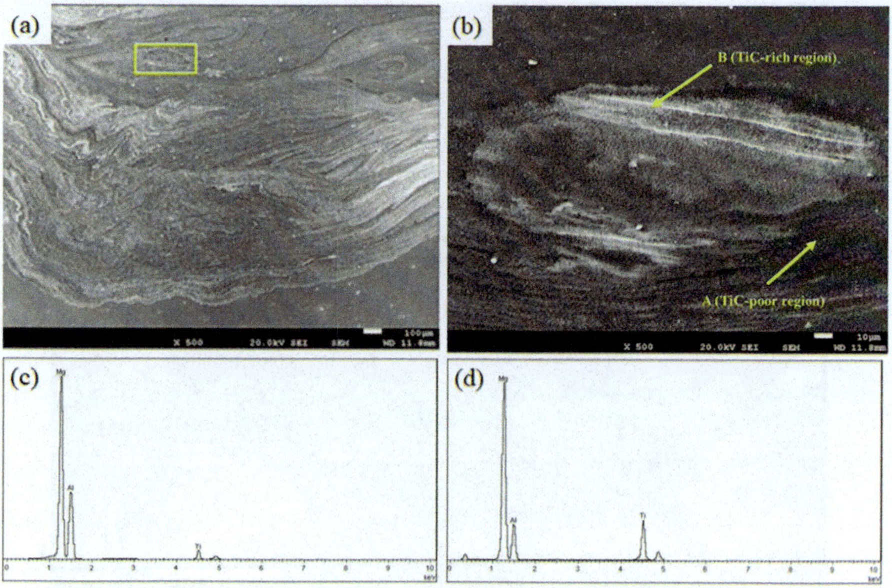

Fig. 5.5 FESEM image of FSW specimen with TiC particles: (**a**) cross-sectional macrostructure, (**b**) high magnification of distinguished zone in SZ, EDS of (**c**) TiC-poor and (**d**) TiC-rich areas from Regions A and B, respectively

Table 5.3 EDS analysis for the samples with reinforcing particles

Element	Mg		Al		Ti		Other	
Region	wt. %		wt. %		wt. %		wt. %	
A	67.6		23.3		5.2		3.9	
B	58.3		17.7		20.4		3.6	

Fig. 5.7c, the distribution of TiC nanoparticles appears more homogeneous in the interface, attributed to desirable heat input and stirring action in the SZ. A higher homogeneous distribution of TiC nanoparticles enhances strengthening mechanisms such as the pinning effect. Conversely, agglomerated TiC nanoparticles and their uneven distribution in various areas of the SZ limit the effectiveness of the indicated strengthening mechanism in developing joint strength.

5.3.2 *Mechanical Properties*

The mechanical characteristics of FSW samples are influenced by several microstructure factors, including grain size, dislocation movement or dislocation density, and the interaction between reinforcing particles and the base material. As per the well-established Hall–Petch relationship described in Eq. 5.1, a decrease in grain size leads to an increase in tensile strength [34].

Fig. 5.6 High-magnification FESEM micrograph of TiC distribution in Region B

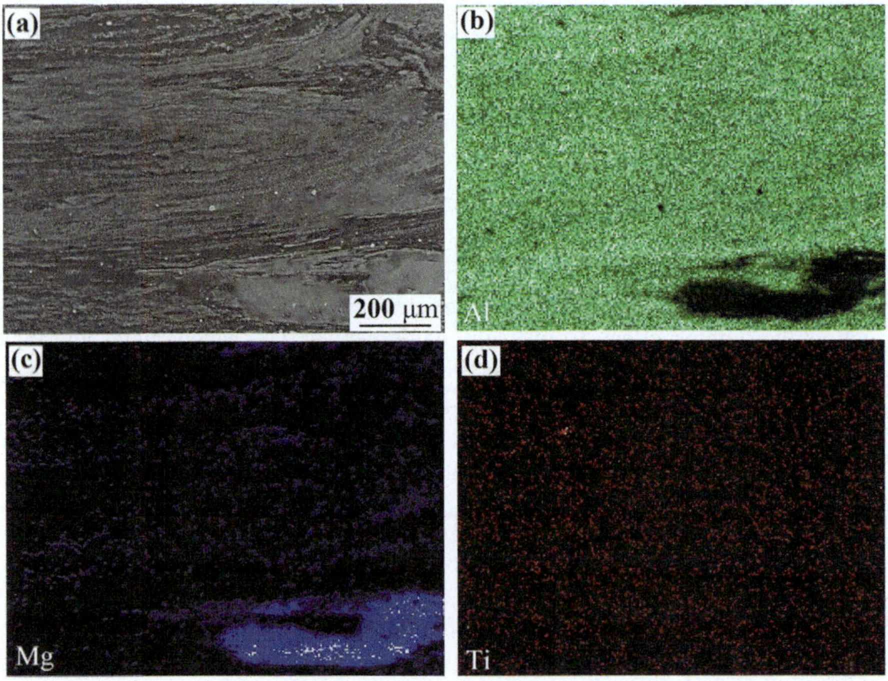

Fig. 5.7 (**a**) FESEM micrograph of SZ at weld interface and EDS mapping for (**b**) Al, (**c**) Mg, and (**d**) Ti

$$\sigma = \sigma_i + k\, D^{1/2} \tag{5.1}$$

In the Hall–Petch relationship, where σ represents the yield strength, σ_i is related to friction stress, k expresses the material constant parameter, and D denotes the grain size. Grain size fundamentally governs the mechanical characteristics of joint samples during conventional FSW processes. In addition, other factors such as the distribution of particles, bonding between the matrix and reinforcements, and dislocations formed due to the unequal thermal expansion coefficient of reinforcement particles and the matrix play significant roles in influencing the mechanical properties.

The strengthening potential of a composite depends on its ability to transfer stress from the matrix to the reinforcement particles. Therefore, achieving a robust interfacial bond between the reinforcing particles and the matrix is crucial. If the interfacial bond is unstable or weak, the interface may fail before effective stress transfer to the particle can occur [35].

5.3.2.1 Tensile Properties

In many engineering applications, achieving increased strength and toughness is essential. Grain refinement through the generation of second phases has long been a crucial method to enhance both toughness and strength in metallurgical processes. However, various factors can adversely affect the microstructure of the weld zone, consequently impacting weld joint strength. These factors include inappropriate heat input; material flow; agglomeration of reinforcing nanoparticles due to weak metallurgical bonding and mechanical interlocking; internal defects such as cavities, cracks, and microvoids; and notably, the formation of IMCs. Such unfavorable conditions create sites susceptible to stress concentration, leading to the initiation and propagation of cracks and resulting in brittle fracture.

Figure 5.8 shows the fracture locations of tensile samples corresponding to weld specimens fabricated with reinforcing particles. According to X-ray diffraction (XRD) analysis patterns from the interface of dissimilar magnesium and aluminum alloy joints reported by Sharifi et al. [17], the presence of IMC layers at the interface joint was confirmed, and fractures were found to initiate and propagate along the hard and brittle IMC layers. In addition, they observed that the tensile direction of the sample is perpendicular to the direction of the TMAZ and the SZ boundary, creating a weak zone. These defects lead to a decrease in the loading region, resulting in stress concentration, crack initiation at defects, propagation along the joint interface, and ultimately brittle fracture of the weld joint.

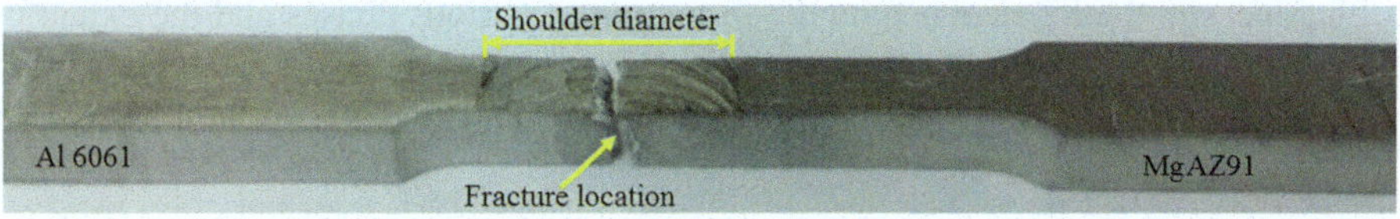

Fig. 5.8 Fracture location in the tensile specimen

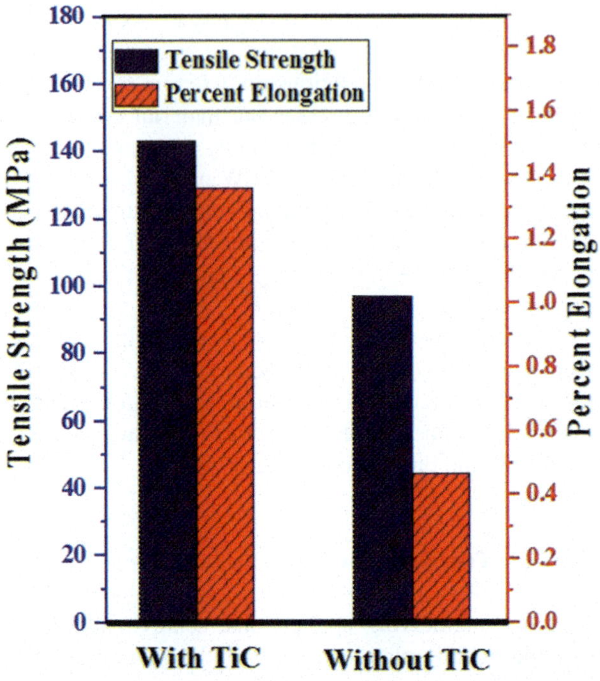

Fig. 5.9 Tensile testing results for the samples

As shown in Fig. 5.9, the highest tensile strength of weld specimens is observed in weld samples fabricated with reinforcing particles. Similar investigations have shown that the introduction of nanoparticles results in grain refinement and enhances the strength of the weld sample by pinning the grain boundaries and nucleation sites [28]. These findings are likely attributed to several factors, including adequate homogeneous nanoparticle distribution, complete intermixing of materials, and the absence of cracks and cavities. These conditions contribute to an increase in the effective working area, a reduction in stress concentration, and consequently a higher tolerance to tensile stress.

5.3.2.2 Hardness

It is important to note that reinforcing particles contribute to the increased hardness level of weld specimens through two mechanisms. First, nanoparticles reduce grain size, leading to increased hardness. In addition, the inherently ultra-hard nature of reinforcing particles further enhances the hardness of the joint [36]. The distribution of hardness in the SZ depends on the formation of various phases and the extent of recrystallization, resulting in the generation of very fine grains and an improvement in dislocation density. Peaks in hardness are often associated with the presence of

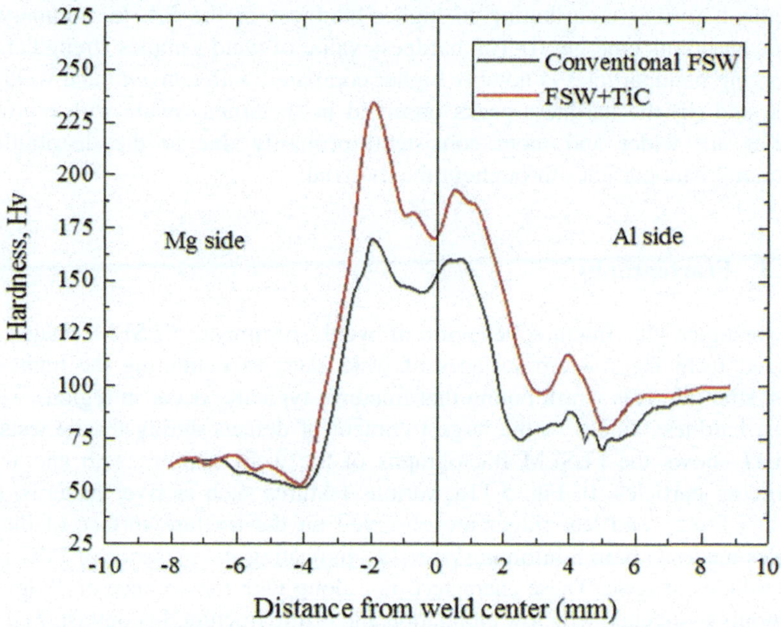

Fig. 5.10 Microhardness variation along the centerline of the cross-section of the samples with and without reinforcing particles

hard and brittle IMCs, with their abundance directly influencing hardness levels. Moreover, variations in hardness across the SZ can be attributed to intercalated banded structures [37].

Figure 5.10 shows the hardness variation of weld samples with and without reinforcing incorporation. It is evident that the hardness value significantly decreases on the aluminum side. This reduction can be attributed to the dependence of precipitation-hardened aluminum alloys, such as those in the 6000 series, on factors such as the size and separation of strengthening precipitates, the degree of dynamic recrystallization, and initially the morphologies of the IMCs, rather than solely on grain size [38]. Consequently, the further reduction of hardness in these areas may be associated with the dissolution, reprecipitation, and coarsening of precipitates due to the heating cycle and high peak temperature. On the other hand, this behavior is not observed on the magnesium side, as AZ91 alloy is primarily hardened by a solid solution [8].

The addition of TiC reinforcing particles leads to an increase in the density of dislocations due to the thermal mismatch between the reinforcing nanoparticles and the matrix. Moreover, the hard nature of the reinforcements and their resistance to plastic deformation during indentation result in a significant increase in the hardness of joint specimens, approximately twice as much compared to conventional weld specimens. In addition, the microstructure of the reinforced weld specimen, with its finer grain size as discussed earlier, enhances resistance against dislocation movement, further contributing to higher hardness in the SZ. In summary, two significant points emerge: (i) The hardness value of weld samples strengthened by reinforcing nanoparticles is notably higher compared with conventional weld specimens, and (ii) the hardness peaks observed in specimens made with reinforcing particles are wider and more consistent, primarily due to the distribution of reinforcing nanoparticles throughout the material.

5.3.2.3 Fractography

To investigate the fracture behavior of weld specimens, FESEM images were analyzed from the fracture surfaces of weld samples exhibiting the highest and lowest strength. It is worth noting that fractures typically occur in regions with the highest hardness values or the largest content of defects during tensile tests. Figure 5.11 shows the FESEM fractographs of the weld samples with and without reinforcing particles. In Fig. 5.11a, various features such as river patterns, quasi-cleavage facets, and tear ridges are observed on the fracture surface of the weld samples made without reinforcing particles, indicating the presence of IMC formation at the weld zone. These characteristics, along with the absence of dimples, are commonly associated with low elongation and brittle fracture. In contrast, Fig. 5.11b shows a relatively varied fracture mode for weld samples with reinforcing particles, with the presence of microvoids and dimples indicating higher ductility. These results can be attributed to factors such as high grain refinement, homogeneous distribution of TiC nanoparticles, and their presence on the fracture surface, all contributing to the strengthening of the weld joint.

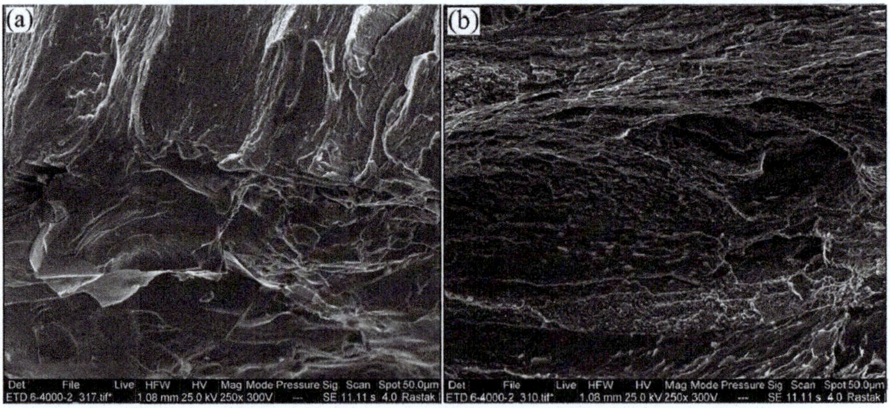

Fig. 5.11 FESEM fractograph of the welds (**a**) without and (**b**) with TiC particles

5.4 Conclusions

In this study, FSW of dissimilar AZ91 to AA6061 was successfully conducted by incorporating TiC nanoparticles. A novel groove design was introduced to embed the reinforcing TiC particles without any dissipation during the welding process. It is important to note that pinning mechanisms and dynamic recrystallization significantly influenced the microstructure evolution and the development of mechanical properties. However, the formation of IMCs in the weld zone had a negative and critical impact on joint sample strength. The key findings of this study are as follows:

- Weld joints were fabricated with and without TiC-reinforced powders via the FSW method, resulting in sound weld joints without apparent defects.
- The incorporation of TiC reinforcement led to a maximum strength due to optimal heat input and uniform distribution of reinforcement particles.
- The addition of TiC nanoparticles notably enhanced the hardness profiles. The EDS analysis and FESEM images revealed uniform distribution of these particles in some areas, while agglomeration was observed in others.
- The FESEM images of the fracture surface of the joints showed brittle fracture behavior for the weld samples made without TiC reinforcement, whereas samples incorporating TiC nanoparticles exhibited microvoids and features, indicating a relatively ductile mode fractures in different locations.

References

1. You S, Huang Y, Kainer KU, Hort N. Recent research and developments on wrought magnesium alloys. J Magn Alloys. 2017;5(3):239–53.
2. Xu RZ, Ni DR, Yang Q, Liu Z, Ma ZY. Influence of Zn interlayer addition on microstructure and mechanical properties of friction stir welded AZ31 Mg alloy. J Mater Sci. 2015;50(12): 4160–73.
3. Abbasi M, Bagheri B, Sharifi F. Simulation and experimental study of dynamic recrystallization process during friction stir vibration welding of magnesium alloys. Trans Nonferrous Metals Soc China. 2021;31:2626–50. https://doi.org/10.1016/S1003-6326(21)65681-9.
4. Hejazi I, Mirsalehi SE. Effect of pin penetration depth on double-sided friction stir welded joints of AA6061-T913 alloy. Trans Nonferrous Met Soc China. 2016;26:676–83.
5. Gupta M, Wong W. Magnesium-based nanocomposites: lightweight materials of the future. Mater Charact. 2015;105:30–46.
6. Lakshminarayanan AK, Annamalai VE. Fabrication and performance evaluation of dissimilar magnesium aluminium alloy multi-seam friction stir clad joints. Trans Nonferrous Metals Soc China. 2017;27(1):25–35.
7. Abdollahzadeh A, Shokuhfar A, Cabrera JM, Zhilyaev AP, Omidvar H. In-situ nanocomposite in friction stir welding of 6061-T6 Aluminum alloy to AZ31 magnesium alloy. J Mater Process Tech. 2018; https://doi.org/10.1016/j.jmatprotec.2018.08.025.
8. Fu B, Qin G, Li F, Meng X, Zhang J, Wu C. Friction stir welding process of dissimilar metals of 6061-T6 aluminum alloy to AZ31B magnesium alloy. J Mater Process Technol. 2015;218:38–47.

9. Zhao Y, Luo Y, Luo Y, He Y, Guo X, Wang S, Cui H, Zhang Y, Wang Z. Effect of welding parameters on the microstructures and mechanical properties of double-pass aluminum/magnesium dissimilar metal friction stir lap welding joint. Mater Today Commun. 2021;26:102132. https://doi.org/10.1016/j.mtcomm.2021.102132.
10. Ding Z, Li Z, Li H, Chen Y. Microstructure of Mg solid solution layer during multi-pass FSP of Mg/Al Composite Plates. Vacuum. 2020;172:109078. https://doi.org/10.1016/j.vacuum.2019.109078.
11. Yin F, Liu C, Zhang Y, Qin Y, Liu N. Effect of Ni interlayer on characteristics of diffusion bonded Mg/Al joints. Mater Sci Technol. 2018;34(9):1104–11. https://doi.org/10.1080/02670836.2018.1424382.
12. Abdollahzadeh A, Bagheri B, Vaneghi AH, Shamsipur A, Mirsalehi SE. Advances in simulation and experimental study on intermetallic formation and thermomechanical evolution of Al-Cu composite with Zn interlayer: effect of spot pass and shoulder diameter during the P-FSSW process. Proc Inst Mech Eng Part L J Mater Des Appl. 2022;237:1–20. https://doi.org/10.1177/14644207221146981.
13. Vaneghi AH, Bagheri B, Shamsipur A, Mirsalehi SE, Abdollahzadeh A. Investigations into the formation of intermetallic compounds during pinless friction stir spot welding of AA2024-Zn-pure copper dissimilar joints. Weld World. 2022;66(11):2351–69. https://doi.org/10.1007/s40194-022-01366-6.
14. Aallahi AA, Shokuhfar A, Moghaddam AO, Abdolahzadeh A. Analysis of SiC nano-powder effects dissimilar on friction stir welding of Al-Mg alloy to A316L stainless steel. J Manuf Process. 2017;30:418–30.
15. Moradi MM, Jamshidi Aval H, Jamaati R. Microstructure and mechanical properties in nano and microscale SiC-included dissimilar friction stir welding of AA6061-AA2024. Mater Sci Technol. 2017;32:1–14.
16. Tabasi M, Farahani M, Givi MB, Farzami M, Moharami A. Dissimilar friction stir welding of 7075 aluminum alloy to AZ31 magnesium alloy using SiC nanoparticles. Int J Adv Manuf Technol. 2016;86:705–15.
17. Asl NS, Mirsalehi SE, Dehghani K. Effect of TiO2 nanoparticles addition on microstructure and mechanical properties of dissimilar friction stir welded AA6063-T4 aluminum alloy and AZ31B-O magnesium alloy. J Manuf Process. 2019;38:338–54. https://doi.org/10.1016/j.jmapro.2019.01.023.
18. Rao H, Ghaffari B, Yuan W, Jordon J, Badarinarayan H. Effect of process parameters on microstructure and mechanical behaviors of friction stir linear welded aluminum to magnesium. Mater Sci Eng A. 2016;651:27–36.
19. Yong Y, Zhang D-T, Cheng Q, Zhang W. Dissimilar friction stir welding between 5052 aluminum alloy and AZ31 magnesium alloy. Trans Nonferrous Met Soc China. 2010;20:619–23.
20. Bagheri B, Abdollahzadeh A, Sharifi F, Abbasi M. The role of vibration and pass number on microstructure and mechanical properties of AZ91/SiC composite layer during friction stir processing. Proc Inst Mech Eng C J Mech Eng Sci. 2022;236(5):2312–26. https://doi.org/10.1177/09544062211024281.
21. Ji S, Meng X, Liu Z, Huang R, Li Z. Dissimilar friction stir welding of 6061 aluminum alloy and AZ31 magnesium alloy assisted with ultrasonic. Mater Lett. 2017;201:173–6.
22. Meng X, Jin Y, Ji S, Yan D. Improving friction stir weldability of Al/Mg alloys via ultrasonically diminishing pin adhesion. J Mater Sci Technol. 2018;34:1817.
23. Bagheri B, Abbasi M, Sharifi F, Abdollahzadeh A. A different attempt to improve friction stir brazing: effect of mechanical vibration and rotational speed. Met Mater Int. 2021;28(9):2239–51. https://doi.org/10.1007/s12540-021-01121-4.
24. Huang Y, Meng X, Xie Y, Wan L, Lv Z, Cao J, et al. Friction stir welding/processing of polymers and polymer matrix composites. Compos Part A Appl Sci Manuf. 2018;105:235–57.

25. Zhao Y, Lu Z, Yan K, Huang L. Microstructural characterizations and mechanical properties in underwater friction stir welding of aluminum and magnesium dissimilar alloys. Mater Des. 2015;65:675–81.
26. Abdollahzadeh A, Bagheri B, Shamsipur A. Development of Al5083/Cu/SiC bimetallic nanocomposite by friction stir spot welding. Mater Manuf Process. 2022;38:1–11. https://doi.org/10.1080/10426914.2022.2157435.
27. Azizieh M, Alavijeh AS, Abbasi M, Balak Z, Kim H. Mechanical properties and microstructural evaluation of AA1100 to AZ31 dissimilar friction stir welds. Mater Chem Phys. 2016;170:251–60.
28. Mofid M, Abdollah-Zadeh A, Gür CH. Investigating the formation of intermetallic compounds during friction stir welding of magnesium alloy to aluminum alloy in air and under liquid nitrogen. Int J Adv Manuf Technol. 2014;71:1493–9.
29. Jamalian HM, Ramezani H, Ghobadi H, Ansari M, Yari S, Givi MKB. Processing–structure–property correlation in nano-SiC-reinforced friction stir welded aluminum joints. J Manuf Process. 2016;21:180–9.
30. Bagheri B, Abdollahzadeh A, Shamsipur A. A different attempt to analysis friction stir spot welding of AA5083-copper alloys. Mater Sci Technol. 2022;39:1–11. https://doi.org/10.1080/02670836.2022.2159633.
31. Mirjavadi SS, Alipour M, Emamian S, Kord S, Hamouda A, Koppad PG, et al. Influence of TiO2 nanoparticles incorporation to friction stir welded 5083 aluminum alloy on the microstructure, mechanical properties and wear resistance. J Alloys Compd. 2017;712:795–803.
32. Bagheri B, Alizadeh M, Mirsalehi SE, Shamsipur A, Abdollahzadeh A. Nanoparticles addition in AA2024 aluminum/pure copper plate: FSSW approach, microstructure evolution, texture study, and mechanical properties. JOM. 2022;74(11):4420–33. https://doi.org/10.1007/s11837-022-05481-z.
33. Bahrami M, Dehghani K, Givi MKB. A novel approach to develop aluminum matrix nanocomposite employing friction stir welding technique. Mater Des. 2014;53:217–25.
34. Abdollahzadeh A, Shokuhfar A, Cabrera J, Zhilyaev AP, Omidvar H. In-situ nanocomposite in friction stir welding of 6061-T6 aluminum alloy to AZ31 magnesium alloy. J Mater Process Technol. 2019;263:296–307.
35. Gao Y, Morisada Y, Fujii H, Liao J. Dissimilar friction stir lap welding of magnesium to aluminum using plasma electrolytic oxidation interlayer. Mater Sci Eng A. 2018;711:109–18.
36. Byung-Wook A, Don-Hyun C, Yong-Hwan K, Seung-Boo J. Fabrication of SiCp/AA5083 composite via friction stir welding. Trans Nonferrous Met Soc China. 2012;22:634–8.
37. Masoudian A, Tahaei A, Shakiba A, Sharifianjazi F, Mohandesi JA. Microstructure and mechanical properties of friction stir weld of dissimilar AZ31-O magnesium alloy to 6061-T6 aluminum alloy. Trans Nonferrous Met Soc China. 2014;24:1317–22.
38. Liu Z, Ji S, Meng X. Joining of magnesium and aluminum alloys via ultrasonic assisted friction stir welding at low temperature. Int J Adv Manuf Technol. 2018:1–10.

Chapter 6
Post-weld Heat Treatment and Nitrogen Application in Welding of Super Duplex Stainless Steels

Mahmoud Abbasi, Ali Tahaei, and Behrouz Bagheri Vanani

6.1 Introduction

Super duplex stainless steels (SDSS) are widely used in various industries, including aerospace, construction, chemical, marine, petrochemical, offshore oil, and gas, owing to their dual-phase and heterogeneous structure comprising an equivalent portion of austenite and ferrite phases [1]. Compared with other stainless steels, SDSS exhibits higher yield strength, ductility, and toughness, while demonstrating intermediate behavior in thermal expansion coefficient and heat transfer characteristics among its constituents [2]. However, the increased proportions of molybdenum and chromium in these materials lead to low localized resistance to corrosion, necessitating a balance between the austenite and ferrite phases in SDSS alloy structures [3].

Various joining methods including welding are commonly employed for SDSS alloys in most applications, thereby affecting the characteristics of materials near the welded areas due to the heating cycle during the welding process [4]. Determining various parameters influencing the weldability of SDSS alloys is crucial for practical utilization, as joining methods significantly impact these materials. Apart from phase balance, a fundamental concern in SDSS is the generation of intermetallic components (IMCs) and other harmful phases during the joining process, which control the mechanical performance and corrosion resistance of joint specimens [5, 6]. Studies by Pramanik et al. [7] focused on various joining processes for the weldability of SDSS alloys, revealing that discrepancies in thermal cycle and heat input during

M. Abbasi · B. B. Vanani (✉)
Department of Materials and Metallurgical Engineering, Amirkabir University of Technology, Tehran, Iran
e-mail: b.bagheri@aut.ac.ir

A. Tahaei
Kharazmi University, Tehran, Iran

S. Gürgen (ed.), *Advances in Manufacturing for Aerospace Alloys*, Sustainable Aviation, https://doi.org/10.1007/978-3-031-64455-9_6

different joining methods affect the microstructure characteristics and functional behaviors of SDSS weld samples. Khan et al. [8] examined the role of intermetallic and secondary phases on the functional properties of SDSS alloys, demonstrating their significant impact on wear performance and the reaction kinetics in acidic environments. Zhang et al. [9] utilized wire and arc additive manufacturing of SDSS to study microstructure evolution and mechanical characteristics, finding that the austenite/ferrite phase balance was disrupted, and the presence of CrN and Cr_2N affected the toughness and ductility of weld samples due to overgrowth of the austenite phase. He et al. [10] investigated the aging process on the formation of intermetallic phases in SDSS alloys, observing the formation of the sigma phase at grain boundaries, which grew with prolonged aging and eventually covered grain boundaries and diffused inside grains.

As previously mentioned, the balance of austenite–ferrite noticeably influences the weld characteristics of SDSS. While the phase balance is achieved in the parent material through a suitable mixture of composition and heat treatment, estimating the austenite–ferrite balance in joint samples is less precise. The primary objective of joining SDSS is to achieve outstanding microstructure evolution and mechanical performance in the fusion zone (FZ) and heat-affected zone (HAZ) for various applications [11]. The microstructure of joint samples depends on the heating and cooling cycles during the joining process, resulting in the formation of transverse and intergranular and intragranular austenite grains along with large ferrite grains in the weld zone [12]. Therefore, proper selection of joint metal constituents, joining parameters, and secondary processes after welding is necessary to achieve weld samples with suitable characteristics for various applications.

Amiri et al. [13] utilized various filler metals for the gas tungsten arc welding (GTAW) of S32750 joints, observing that the extended ferrite–austenitic structure converted into a dendritic γ-austenite structure within the α-ferrite matrix. Migiakis et al. [14] employed a mixture of argon and nitrogen in plasma and shielding gas during the welding of SDSS alloys, noting positive effects on the mechanical characteristics of weld specimens. The use of nitrogen in the shielding gas enhanced the nitrogen content and austenite generation of the joint metal for single-pass and multi-pass joints in both as-welded and heat-treated conditions during the GTAW process [15]. Bermejo et al. [16] investigated the role of shielding gases on the microstructure and mechanical characteristics of SDSS joints fabricated by the gas metal arc welding method, finding that shielding gases containing 30% helium provided excellent results in terms of high ductility and impact toughness. Post-weld heat treatment (PWHT) was suggested to address concerns related to excessive ferrite content and intermetallic phases. Ferreira et al. [17] applied solution heat treatment for joining S31803 duplex stainless steel, concluding that treatment at 1150 °C for 10 s removed chromium nitrides and reestablished the balanced ferrite/austenite phase ratio. Zhang et al. [18] examined the effects of electron beam welding and PWHT at various temperatures on the microstructure evolution of SDSS alloys, finding that the optimized solution temperature range between 1050 and 1100 °C led to austenite generation, dissolution of Cr_2N, and elimination of dendritic segregation.

The discussed points underscore the importance of SDSS alloys and the complexity involved in maintaining their characteristics after the joining process. The variation in structure and mechanical performance is a nuanced process, contingent upon factors such as welding conditions, filler metals, and secondary treatments. However, there has been no investigation into the effects of nitrogen gas existence and the application of PWHT in GTAW of SDSS. Given the critical role of phase balance in the weld characteristics, this study aims to examine the effects of nitrogen gas application and PWHT on the structure and mechanical properties of S32750 joints produced via the GTAW process.

6.2 Experimental Details

In the present study, GTAW technique was employed to weld an S32750 sheet with dimensions of 50 × 50 × 3 mm. Table 6.1 presents the chemical compositions of both the base and filler materials.

The gap between the sheets to be welded was cleaned with alcohol to remove all dust and contamination. Joining parameters were determined based on Table 6.2. Figure 6.1 shows a schematic design using a mixture of argon and nitrogen as shielding gas for welding. It is worth mentioning that the nitrogen evolution occurred at the joint pool, and no bubbles or porosity was created in the joint area due to optimized welding conditions.

The process involved the addition of nitrogen gas along with existing argon as a shielding gas. PWHT was performed at 1100 °C for 10 min, followed by water quenching. This temperature was chosen as it is higher than the solubilization temperature of detrimental intermetallic formations, allowing for the attainment of weld specimens with high toughness. Table 6.3 indicates the specimens and their processing conditions used in this study.

The microstructure investigation was conducted following the ASTM E3 standard. After cutting the cross-section of the weld joints and mounting, a standard metallography procedure was followed, which involved grinding with a series of water-cooled SiC emery papers ranging from 400 to 1200 grit size, followed by polishing with 6 and 3 μm Al_2O_3. The Beraha's reagent (50 mL H_2O, 50 mL HCl,

Table 6.1 Chemical compositions of the base and filler materials

Element (wt.%)	C	Cr	Ni	Mo	Mn	Si	Cu	N	Fe
Base: S32750	<0.03	24–26	6–8	3.5–4.5	<1.2	<0.08	<0.05	0.24–0.32	Bal.
Filler: ER2209	0.01	22.8	8–9	3.1	1.6	0.4	0.16	0.16	Bal.

Table 6.2 Welding parameters

Current (A)	Voltage (V)	Travel speed (mm/min)	Heat input (kJ/mm)	Shielding gas	Gas flow rate (L/min)
80	12–15	50–70	1.14	Ar 99.99%	12

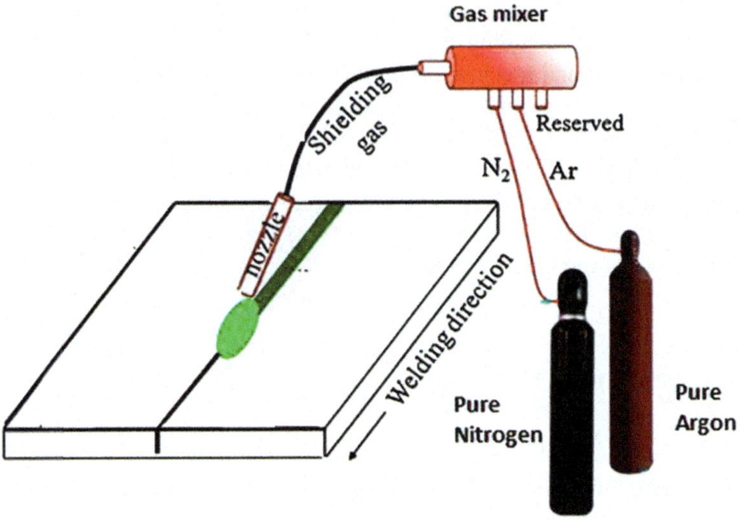

Fig. 6.1 Schematic illustration of the welding process

Table 6.3 Processing conditions for the specimens

Specimen code	Condition
SFN2	Welding with filler metal + addition of nitrogen
SFN2P	Welding with filler metal + addition of nitrogen and PWHT

1 g $K_2S_2O_5$, and 10 g $NH_4F.HF$) was applied to etch and reveal the microstructures in the specimens. Due to the high percentage of alloying elements in SDSS, electrochemical etching was performed using a power supply equipment model PS-305D with 40% NaOH electrolyte at 20 V for 15 s, in addition to chemical reagent etching. In this method, a stainless steel rod served as the cathode, while the sample acted as the anode. Figure 6.2 illustrates the setup for electrochemical etching. Metallography investigations were conducted using a Leica MEF4M optical microscope, and the phase percentages were estimated using the Image-Pro Plus software.

The phase characterization of the weld joints was analyzed using X-ray diffraction (XRD) analysis. The Bruker AXS D-8 XRD equipment was employed, utilizing CuKα radiation ($\lambda = 0.154$ nm) at 40 kV and 20 mA. The XRD spectra of the specimens were recorded in the 2θ range between 10° and 110° with an angular step of 0.01° and a counting time of 2 s. The evaluation of the peaks for detecting the existing phases was carried out using the DIFFRAC PLUS evaluation software.

To analyze the hardness profile of the weld specimens, Vickers microhardness measurements were conducted on the base metal (BM), FZ, and HAZ. A Future-Tech microhardness machine, applying a load of 1 kg for 15 s, was utilized for hardness measurements. Tensile tests were performed on the specimens according to ASTM E8M standards using a Tinius Olsen 60 machine. Following the tensile test,

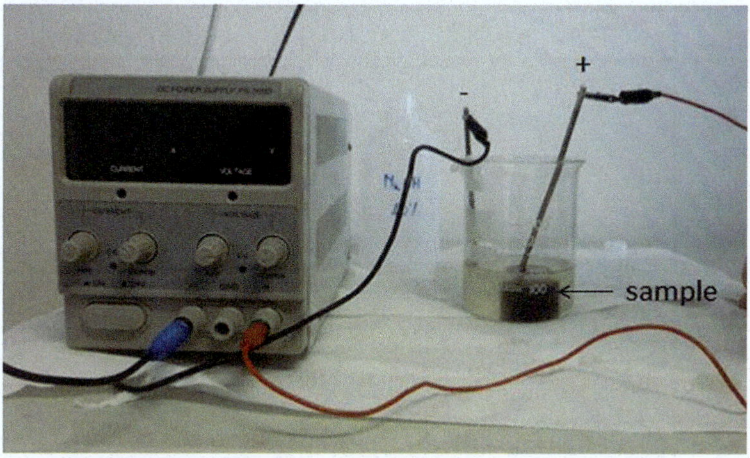

Fig. 6.2 Electrochemical etching setup

the fracture surfaces of the specimens were investigated using a stereomicroscope (Leica MZ6) and a scanning electron microscope (SEM; Zeiss EVO EMA 15). Prior to the surface examination, all specimens underwent ultrasonic cleaning in an acetone solution for approximately 5 min to remove dust and contaminants from the fracture surfaces.

6.3 Results and Discussion

6.3.1 Microstructural Properties

Figure 6.3 shows the microstructure of the specimen with the nitrogen application and subsequent PWHT. The microstructure in the HAZ comprises intergranular austenite islands dispersed within a ferrite matrix. Following PWHT, the structure remains largely unchanged, with an increase in the size of austenite laths and globularity attributed to solution annealing at elevated temperatures. In the FZ of this specimen, the presence of nitrogen as a strong austenite stabilizer results in a microstructure consisting of complete austenite, with only a few small ferrite islands remaining. After welding, the cooling rate emerges as a significant parameter influencing phase transformation in the FZ and HAZ [19]. The slower cooling rate, due to high heat input, promotes increased transformation of ferrite into austenite, leading to a favorable phase balance. Conversely, higher heat input can result in coarsening ferrite grains and precipitation of brittle intermetallic compounds. Conversely, rapid cooling, due to low heat input, leads to increased ferrite content and precipitation of chromium-rich nitrides. Owing to the high percentage of

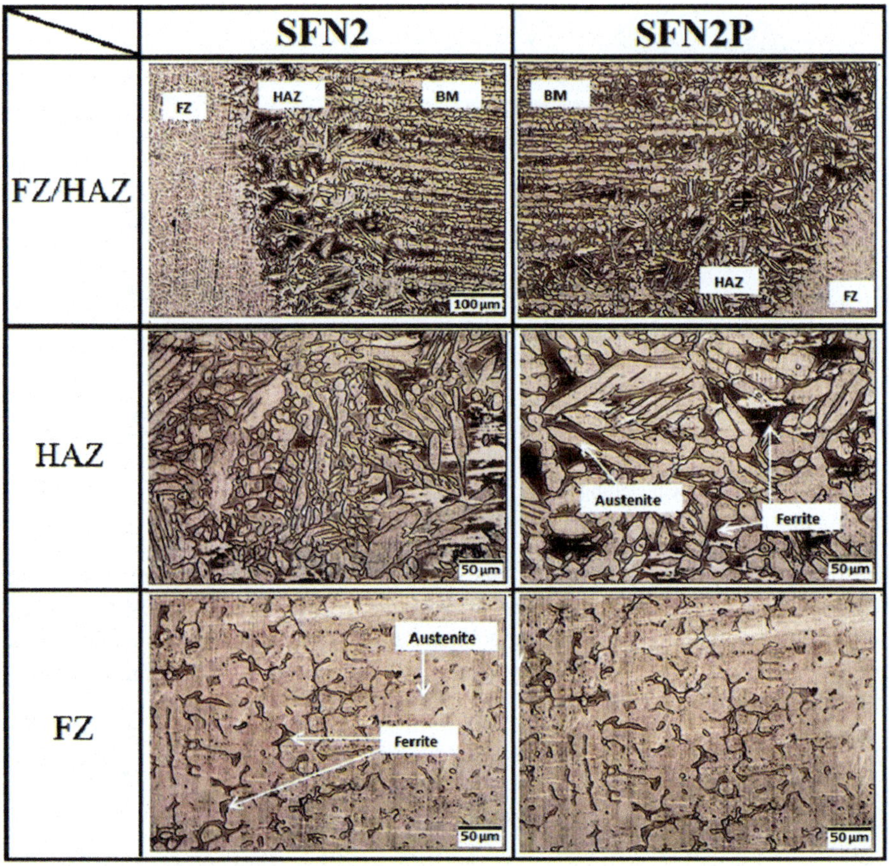

Fig. 6.3 Microstructures of weld zones in the specimens etched by reagent

alloying elements such as nickel and nitrogen, chemical reagent etching was not successful in revealing the microstructure. Consequently, electrochemical etching was performed for better observation. Figure 6.4 shows the microstructure of the FZ and HAZ after electrochemical etching for the specimens.

Differences in the percentages of austenite and ferrite in different welds can be explained by referring to the ternary Fe–Cr–Ni diagram. According to this diagram, solidification primarily commences from delta ferrite and persists throughout the solidification process. The proportion of austenite is primarily influenced by the cooling rate [20]. During solid-state cooling, at temperatures below the ferrite solvus, partial transformation of ferrite to austenite occurs, resulting in relatively small austenite size and content. The sequence of ferrite to austenite formation during cooling rates typically begins with grain boundary allotriomorphs, followed by Widmanstatten side plates and intergranular precipitates. Higher heat input and consequently lower cooling rates result in a greater amount of retained austenite in

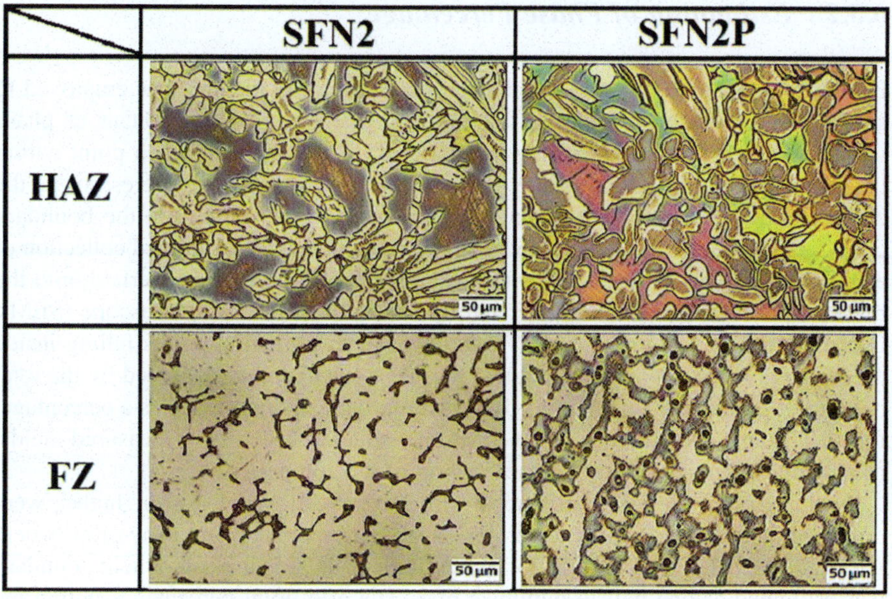

Fig. 6.4 Microstructures of weld zones in the specimens etched by electrolytic etching

the microstructure. Moreover, it is anticipated that using Ar + 2% N_2 during the welding would result in a joint region nitrogen content similar to that of the base material, and higher nitrogen concentrations would enhance the nitrogen content of the joint metal [21]. From the evolution of the microstructure, it can be inferred that an adequate annealing temperature coupled with an appropriate cooling period provides sufficient time for the reformation and growth of austenite, thereby facilitating the attainment of a balanced ratio between the two phases.

The generation of grain boundary and side plate austenite during solidification typically requires reduced driving force and occurs at elevated temperatures with low undercooling. Conversely, the formation of intergranular acicular austenite necessitates greater undercooling and occurs at lower temperatures. It is noteworthy that grain boundary austenite and Widmanstatten side plates form at the onset of high temperatures, whereas intergranular austenite particles require higher driving forces and precipitate later at lower temperatures. At low cooling rates, thin and discontinuous side plate austenite, along with grain boundary and intergranular austenite, are formed. The addition of nitrogen from the base material during the welding process leads to an increased percentage of austenite in the weld zone and reduces the risk of nitride precipitation in ferrite grains. Nitrogen serves two primary roles in SDSS: first, enhancing pitting corrosion resistance by increasing the value of the pitting resistance equivalent number (PREN = % Cr + 3.3% (Mo + 0.5%W) + 16(30) %N) and, secondly, expediting the production of austenite in the weld zone. Regarding the application of PWHT, it is observed that many intergranular austenite grains undergo noticeable growth and primarily assume a spheroidal shape. In addition, austenite grains do not exhibit a preferred growth direction.

6.3.2 Estimation of Phase Percentage

During this technique, a periodic manual point counting method was employed to statistically evaluate the volume fraction of a recognizable component or phase within areas of the structure using a point grid. In this analysis, each point within the examination grid falling inside the structural characteristic of interest, or on the characteristic boundary, was counted as one, with points falling on the boundary considered as one-half. Specifically, an eyepiece reticle with a standard collection of examination points or a transparent plastic examination grid was overlaid onto the micrograph or its projection, generated by a scanning electron microscope (SEM), light microscope, or micrograph. The number of examination points falling inside the phase or the component of interest was then calculated and divided by the total number of grid points, yielding a point fraction, typically expressed as a percentage, for that field. The average point fraction, calculated across n-measured fields, provided an estimate of the volume fraction of the component.

In the present study, micrographs with magnifications of 400 or higher were preferred, and 25 grid points were used on the transparent paper. Any point where uncertainty existed regarding its location within or outside the component boundary was estimated as one-half to minimize bias. The grid point percentages within the component marked on the ith field were calculated using Eq. 6.1.

$$P_{p(i)} = \frac{P_i \times 100}{P_T} \tag{6.1}$$

where P_i represents the point count on the ith field, and P_T is the total number of points in the examination grid. For each zone, a minimum of five images was considered, and finally, the median of percentages was determined as the ferrite percentage of the zone. Therefore, the arithmetic average of $P_{p(i)}$, the standard deviation estimator, s, as well as the 95% confidence interval, 95% CI, must be calculated and evaluated for individual sets of fields. The equations for calculating them are given by Eqs. 6.2, 6.3, and 6.4.

$$P_p = \frac{1}{n}\sum_{i=1}^{n} P_p(i) \tag{6.2}$$

$$s = \left[\frac{1}{n-1}\sum_{i=1}^{n}\left[P_p(i) - P_p\right]^2\right]^{1/2} \tag{6.3}$$

$$95\%\ CI = \frac{t \times s}{\sqrt{n}} \tag{6.4}$$

The solid-state ferrite-to-austenite transformation is governed by the diffusion and redistribution of austenite and ferrite stabilizers for the two phases, and it is incomplete due to the limitation of diffusion by the rapid cooling process. As a

Table 6.4 Phase percentages obtained by manual calculations

Specimen	Austenite	Ferrite
SFN2	68.8	31.2
SFN2P	55.4	44.6

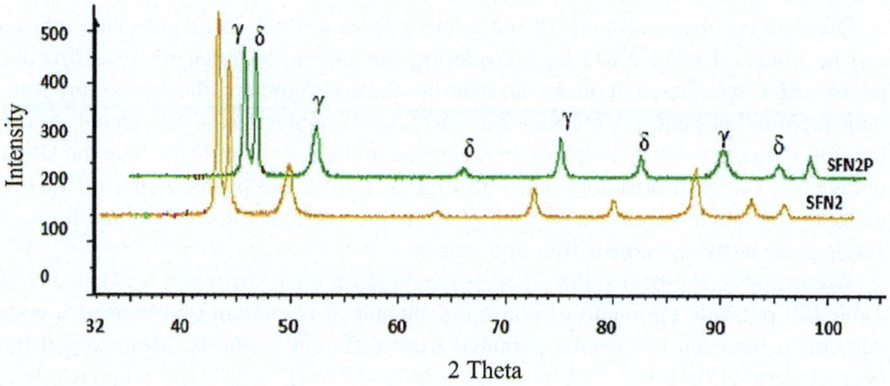

Fig. 6.5 XRD spectra for the specimens

result, only some of the elevated-temperature ferrite is converted into austenite, while the remainder is maintained as skeletal ferrite [22]. The acceptable ferrite-to-austenite ratio and location should be taken into consideration in area measurements. Table 6.4 presents the phase measurements obtained by manual point measurement.

6.3.3 X-Ray Diffraction Analysis

Based on the data obtained from XRD spectra shown in Fig. 6.5, the phases mainly consist of ferrite and austenite with different lattice parameters, as indicated on top of each peak. There is no evidence of undesired intermetallic phases, namely σ, χ, and nitride. Moreover, the morphology and distribution of ferrite and austenite phases in both the joint specimens are similar. This was confirmed by the results obtained from metallography analysis, and it is predicted that further elevated-temperature solution heat treatments will influence the final distribution of austenite and ferrite phases within the microstructure. This suggests that the quality of all the joints was appropriate and the welding parameters were selected correctly. These results are consistent with the findings reported by Zhou et al. [23], who analyzed the impact of various solution annealing heat treatments, ranging from 100 to 1350 °C, on the formation of harmful intermetallic and secondary phases in stainless steels. They concluded that no evident precipitates or tertiary phases were detected in the microstructure. In addition, Kim et al. [24] indicated that the dissolution of Cr_2N is attributed to the diffusion of nitrogen atoms from the ferrite phase to the austenite phase, leading to an improvement in the austenite within the joint metal.

Table 6.5 Phase percentages obtained by XRD results

Specimen	Austenite	Ferrite
SFN2	62	38
SFN2P	54	46

Quantitative measurements of the relative volume fraction of individual phases can be obtained from XRD by considering the integrated intensity of diffraction peaks with respect to each phase. To measure these parameters, the peaks were fitted with a Gaussian peak profile function. Additional analysis was conducted on diffraction patterns to estimate phase percentages in the matrix and calculate the lattice parameters for both austenite and ferrite phases from the peaks of the diffraction pattern. The lattice parameter of each phase is the average value calculated for each single peak in the specific diffraction pattern.

According to the peaks, the phase percentage for each specimen was calculated. Table 6.5 presents the result of phase percentage. It is evident that there is a good agreement between the results obtained from different methods. Moreover, it has been confirmed that the addition of nitrogen and PWHT plays an essential role in reducing element segregation between phases. With the addition of nitrogen, as a quick diffusion interstitial component, the difference between the solidus and formation start temperature of austenite is smaller. As a result, the formation of austenite occurs at temperature values more proximate to the solidus when nitrogen atom diffusions happen more easily, leading to increased portions of austenite formation.

6.3.4 Hardness

Figure 6.6 shows the microhardness profiles for the specimens. As evident, a decrease in hardness value in the FZ was observed due to using filler material with relatively fewer alloying elements compared with the base metal. In addition, the growth of grains in the weld metal due to the thermal cycle can be considered another reason. The decomposition of the ferrite phase, also associated with phase separation (PS), in which the originally homogeneous body-centered cubic (bcc) phase has decomposed into a favorably coherent nanostructure of Fe-rich (α) as well as Cr-rich (α') phases, is an important factor influencing the mechanical characteristics [25]. According to Hosseini et al. [26], the impact of the processing route compared with the variation in composition is more crucial. Higher severe PS of the weld specimens would also be owing to the applied residual strains during the joining process itself, including anticipated increased internal strains increasing PS. It is worth mentioning that austenite may undergo higher expansion compared with the ferrite during the joining temperature degree, and consequently, the ferrite would undergo compressive strains after cooling. Amiri et al. [13] reported the fact that the presence of the α-stabilizer components in the filler rods leads to an improvement in the hardness of the joint metal, while the concentration of γ-austenite, as well as

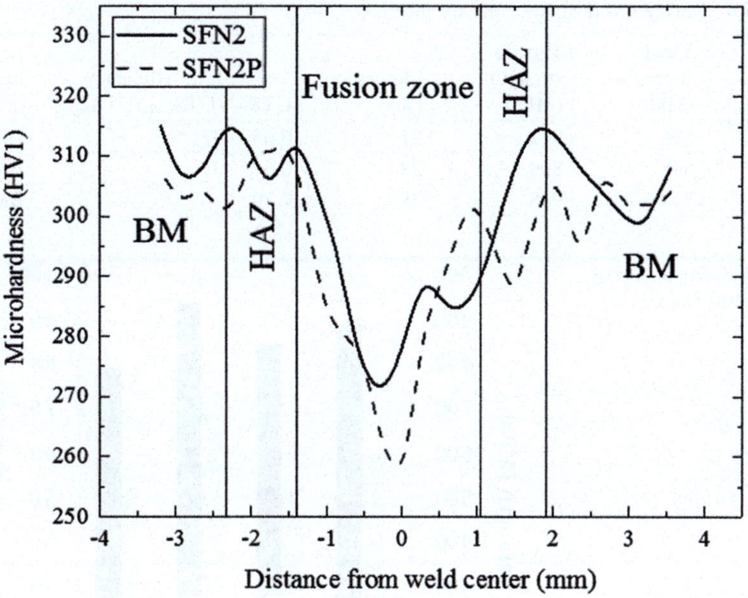

Fig. 6.6 Microhardness profiles for the specimens

δ-ferrite phases in the utilized filler rods, leads to a reduction in the hardness values of the joint material during the joining of SDSS. The authors believe that, at high temperatures, austenite may be converted into ferrite, but the increased nitrogen content stabilizes the austenite phase and enhances the austenite–ferrite ratio in the weld zone. Subsequently, in the ferrite zone, grain growth occurs, and consequently, the ferrite decomposition to austenite is affected by nitrogen diffusion from supersaturated ferrite to reforming austenite during the cooling condition.

Applying PWHT did not show any specific effects on changing the microhardness, and this remained at approximately the same value through the microhardness line profile. Based on the obtained results, it should be noticed that the ferrite phase, in comparison with the austenite phase, shows higher strength behavior at ambient temperature, while at high-temperature values, the austenite phase presents higher high-temperature strength. Therefore, the addition of nitrogen gas, as an austenite stabilizer element, leads to an improvement in the yield strength of the austenite phase, resulting in the transformation from a softer phase to a higher phase.

6.3.5 Tensile Properties

Table 6.6 presents the obtained results from the tensile tests. For easier comparison, Fig. 6.7 shows the bar chart of the tensile results. As mentioned in previous investigations, the addition of various elements during the welding of SDSS is

Table 6.6 Tensile test results for the specimens

Specimen	Yield strength (MPa)	Tensile strength (MPa)	Elongation (%)	YS/ UTS	Joint efficiency (UTS_{weld}/UTS_{base})	Location of fraction
SFN2	596	912	21	0.65	97	HAZ
SFN2P	548	894	28	0.61	95	BM
BM	744	938	28	0.79	–	BM

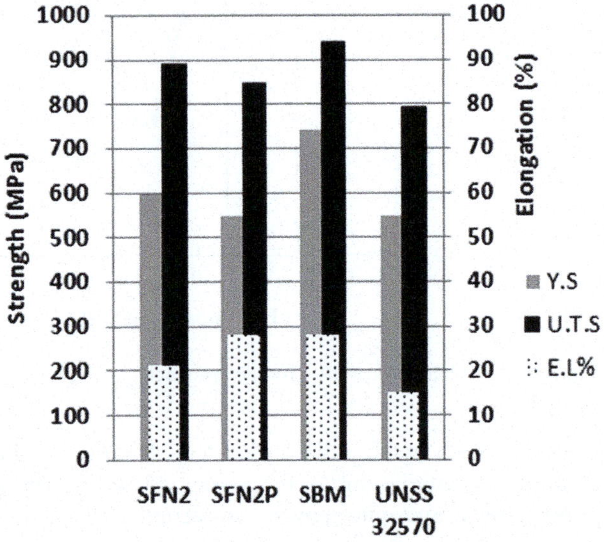

Fig. 6.7 Comparison of tensile test results

dependent on different factors, such as the nature of filler metal, the chemical composition of the base material, heat input or arc energy, and the microstructure condition [27]. Moreover, in the balance between ferrite and austenite phases, cooling rates have been found to play a critical role and consequently affect the mechanical characteristics of weld samples. A high cooling rate causes a higher ferrite content due to the limitation of austenite formation owing to insufficient time and temperature. In addition, the formation of nitrides, because of the diffusion of fewer nitrogen atoms into austenite, is facilitated. Supersaturation of nitrogen, as well as precipitation of nitrides in the ferrite phase, is facilitated as the temperature decreases. On the other hand, higher heating leads to a lower cooling rate, which enhances the production of detrimental intermetallic compounds [28]. As it is clear from the results, after applying PWHT, the value of the yield strength and elongation for the specimen SFN2 decreased by 8%, while the value of elongation increased. The increased strength of specimens before PWHT can be attributed to the existence of an increased percentage of α-ferrite in the weld metal structure. In addition, the austenite phase shows a higher thermal expansion coefficient in all ranges of temperature with a much slower reduction in Young's modulus as a function of temperature. The variant temperature cycles lead to phase-specific residual stresses

[29]. Therefore, with an increase in temperature values, the flow stress of SDSS has been basically near that of ferrite at solution-annealing temperatures at a slow strain rate, whereas at a high strain rate, it is near that of austenite.

After PWHT, microstructural softening occurs to release the internal stresses and decrease dislocation density, resulting in a decrease in tensile strength. Moreover, PWHT is considered a vital procedure for the toughness of weld samples due to its role in restoring phase equilibrium and releasing residual stress. From a microstructure perspective, the grain size of joint metal improves with nitrogen loss. That is, the volume fracture of grain boundaries decreases, and as a result, the density of dislocation decreases. With limitation in the nitrogen loss in the joint metal, the grain size decreases, and the volume fracture of grain boundaries increases, leading to an improvement in ductility.

By evaluating the achieved results from microstructure examination, determination of phase, as well as obtained data from tensile tests, it is no exaggeration to say that the role of PWHT is primarily related to the modification of microstructure in restoring equivalent percentages of austenite and ferrite with no considerable variations in mechanical characteristics. For the evaluation of the tensile test results, two terms should be defined for various industrial applications: The first one is the ratio of yield strength-to-ultimate tensile strength (YS/UTS), where the higher the ratio, the lower the uniform plastic deformation before fracture. Another one is joint efficiency, which can be defined based on the ratio, ultimate tensile strength of weld-to-ultimate tensile strength of base metal (UTS_{weld}/UTS_{base}) [30–32], and 95% represents an acceptable result, which was obtained for all the specimens.

Figure 6.8 shows the macrosection of the tensile test specimens after fracture. Figure 6.9 presents the side-view macrographs of the fractured specimens achieved by a stereomicroscope. The specimens exhibited ductile fracture with a visible necking area. Both specimens of SFN2 were fractured in the HAZ, while the specimen of SFN2P was fractured outside the weld zone. It should be concluded that nitrogen has both positive and negative effects on the mechanical characteristics of welded stainless steels. It affects the austenite-to-ferrite ratio, which can modify the microstructure and cause the formation of nitrides. In addition, it has a

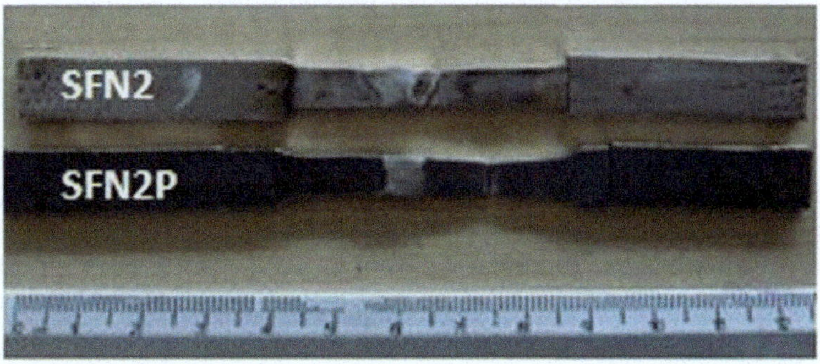

Fig. 6.8 Macroimages of the specimens after tensile testing

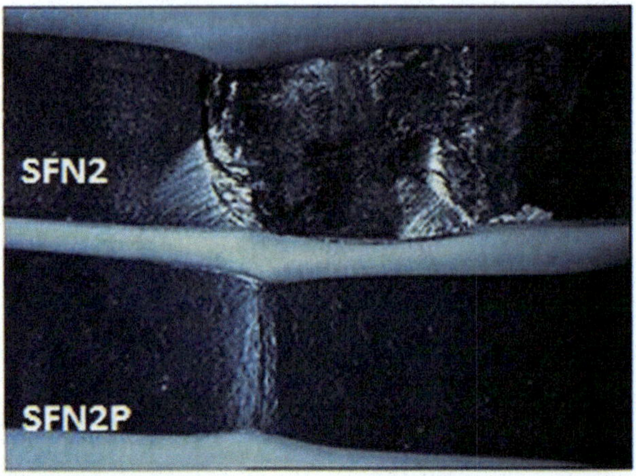

Fig. 6.9 Stereo-macrographs of the fracture surfaces for the specimens

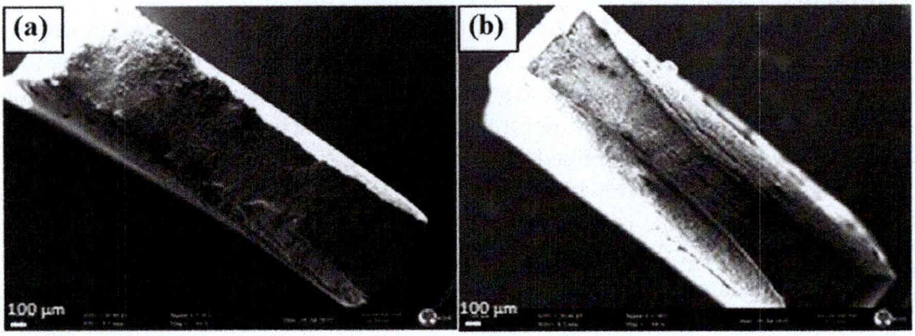

Fig. 6.10 SEM images of the fracture surfaces for (**a**) SFN2 and (**b**) SFN2P

strengthening effect on both the ferrite and austenite phases. Due to the austenite-stabilizing properties of nitrogen, it tends to dissolve in the austenite phase, leading to interstitial solution strengthening in austenite compared with ferrite.

6.3.6 Fracture Surfaces

An SEM investigation was conducted to observe the fracture surface of the specimens as shown in Fig. 6.10. The fracture surface analysis indicates that the specimens with lower elongation correspond to full crack propagation in the HAZ, while higher elongation values are detected in specimens that were broken in the base metal. SFN2P, which was broken in the base metal, indicates a surface that is not very rough, and necking is completely visible. The fracture surface of SFN2, which was broken from the HAZ, also indicates a very rough and inhomogeneous surface without any specifically visible necking. This could be due to the different microstructures in this zone.

Table 6.7 EDS results for the specimens

Element (wt.%)	Cr	Mn	Ni	Si	Mo	Fe	Cr_{eq}	Ni_{eq}	R (Cr_{eq} + Ni_{eq})
SFN2	26.50	1.10	6.38	0.36	4.92	Bal.	31.96	6.93	4.60
SFN2P	28.94	1.51	5.74	0.31	3.34	Bal.	32.28	6.49	4.97

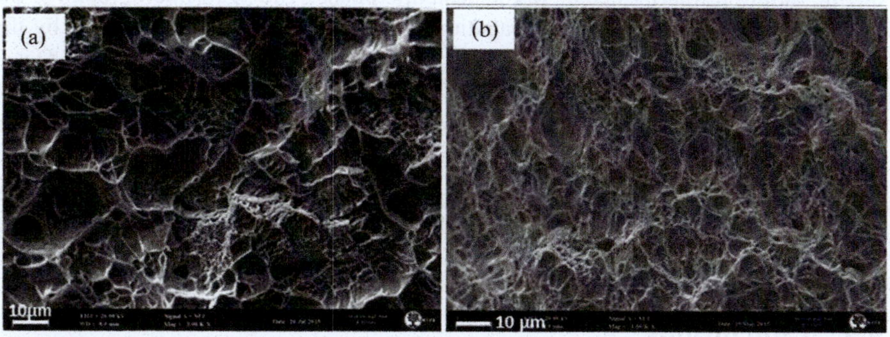

Fig. 6.11 Fracture surfaces in (**a**) SFN2 and (**b**) SFN2P

Table 6.7 indicates the energy-dispersive spectroscopy (EDS) analysis of all fracture surfaces for the specimens under different joining conditions. The measured range of elements calculated by EDS is almost identical to the base material with a litter discrepancy. The chromium contents of the joint metals are higher than that of the base metal. SFN2, which was broken in HAZ, shows little difference in chemical composition. It has been confirmed that nitrogen alloying improves the austenite portion and enhances the partitioning of nickel and nitrogen in the austenite phase and molybdenum and chromium in the ferrite phase [33]. The joining pool size and solidification time have been known as two fundamental variables resulting in nitrogen loss, so increasing nitrogen content may decrease the effects of dendritic segregation of metallic components in the joint. Furthermore, the cooling rate after welding has been recognized as having a significant role in the segregation of alloying elements, which means the lower cooling rate after GTAW can lead to an increase in austenite generation and decrease the effect of dendritic segregation. Based on the obtained data and calculation of *R* value (Cr_{eq}/Ni_{eq} for fracture surfaces), the fracture mainly occurs in the ferrite phase.

As depicted in Fig. 6.11, the presence of dimples across various sizes indicates a predominantly ductile fracture behavior, where plastic deformation occurred before final fracture [34–36]. When overload is the primary cause of fracture, a process known as microcoalescence voids (MCV) governs the failure. Here, microvoids nucleate in regions of localized strain discontinuity, often associated with second-phase particles. Comparing the fracture surfaces of the specimens after PWHT reveals that dimples generally become deeper and wider, suggesting a longer fracture path before final fracture. Notably, SFN2 and SFN2P, where fracture occurred in HAZ and parent metal, respectively, exhibit increased dimple size after heat treatment, possibly due to their higher elongation values. Previous studies [37] suggested that crack initiation occurs in the austenite phase, with propagation typically proceeding through transgranular cleavage of the ferrite phase. The formation of reformed austenite in the joint region, owing to tempering and repetitive heating–cooling cycles in multipass joining, fosters the development of microvoids, favoring ductile failure [38]. Cleavage facets predominantly belong to the ferrite phase, while ductile failure primarily occurs in the austenite phase. The SEM micrographs of the fracture surfaces reveal crack propagation through the base metal, followed by the weld metal comprising austenite and ferrite grains parallel to the surfaces. Occasionally, this propagation transverses with sharper deformation of the austenite phase. The alteration in fracture surface among the specimens may also stem from existing inhomogeneous plastic deformation due to structural changes in the FZ and HAZ. Notably, dimples exhibit elongation along the stress direction, providing insights into the fracture mechanics of the specimens.

6.4 Conclusions

This study investigated the role of nitrogen gas application and PWHT on the GTAW process of S32750. Microstructural analysis reveals that PWHT primarily leads to an increase in the size of austenite laths and globularity, attributed to solution annealing at elevated temperatures. In addition, a balanced austenite/ferrite phase ratio (54:46) is achieved in the specimens treated with nitrogen and PWHT. XRD analysis confirms the presence of ferrite and austenite phases with distinct lattice parameters, and there was no evidence of undesired intermetallic phases. Specimens subjected to both nitrogen and PWHT demonstrate improved mechanical properties, attributed to the restoration of phase equilibrium and release of residual stress. Fracture surface analysis indicates ductile mode failure, with plastic deformation occurring prior to fracture. The study concluded that PWHT significantly influences the microstructural and mechanical characteristics of SDSS weld specimens, ultimately enhancing their functional performance.

References

1. Min W, Baosheng L, Shoulu Z, Guoping L, Lixin W, Lifeng H, Yinghui W. Hot deformation behavior and softening mechanism of as-cast S32750 super duplex stainless steel at low and high strain rates. J Mater Eng Perform. 2020;29:727–38.
2. Li J, Shen W, Lin P, Wang F, Yang Z. Effect of solution treatment temperature on microstructural evolution, precipitation behavior, and comprehensive properties in UNS S32750 super duplex stainless steel. Metal. 2020;10:1481.
3. Paulraj P, Garg R. Effect of welding parameters on pitting behavior of GTAW of DSS and super DSS weldments. Eng Sci Technol. 2016;19:1076–83.

4. Vahman M, Shamanian M, Golozar MA, Jalali A, Sarmadi MA, Kangazian J. The effect of welding heat input on the structure–property relationship of a new grade super duplex stainless steel. Steel Res Int. 2020;91:1900347. https://doi.org/10.1002/srin.201900347.
5. Sato YS, Nelson TW, Sterling CJ, Steel RJ, Pettersson CO. Microstructure and mechanical properties of friction stir welded SAF 2507 super duplex stainless steel. Mater Sci Eng. 2005;397:376–84. https://doi.org/10.1016/j.msea.2005.02.054.
6. De Souza GCH, Da Silva AL, Tavares SS, Pardal JM, Ferreira ML, Filho IC. Mechanical properties and corrosion resistance evaluation of superduplex stainless steel UNS S32760 repaired by GTAW process. Weld Int. 2016;30:432–42.
7. Pramanik A, Littlefair G, Basak AK. Weldability of duplex stainless steel. Mater Manuf Process. 2015;30:1053. https://doi.org/10.1080/10426914.2015.1019126.
8. Khan WN, Chhibber R. Effect of intermetallic and secondary phases on dry and wet sliding wear behavior of super duplex stainless steel. Trib Trans. 2019;63:403.
9. Zhang X, Wang K, Zhou Q, Ding J, Ganguly S, Marzio G, Yang D, Xu X, Dirisu PH, Williams SW. Microstructure and mechanical properties of TOP-TIG-wire and arc additive manufactured super duplex stainless steel (ER2594). Mater Sci Eng A. 2019;762:138097. https://doi.org/10.1016/j.msea.2019.138097.
10. He ML, Yu LM, Li HJ. Evolution of intermetallics in a 26Cr–4Mo–2Ni (Ti + Nb) super ferritic stainless steel aged at 800°C. Mater Res In. 2015;19(1):36–40.
11. Taban E. Joining of duplex stainless steel by plasma arc, TIG, and plasma Arc+ TIG welding processes. Mater Manuf Process. 2008;23(8):871–8.
12. Tathgir S, Bhattacharya A, Bera TK. Influence of current and shielding gas in TiO_2 flux activated TIG welding on different graded steels. Mater Manuf Process. 2014;30:1115.
13. Amiri E, Ostovan F, Toozandehjani M, Shafiei E, Mohamed IF. Study and selection of most appropriate filler rod for GTAW of S32750 super duplex steel joints: a comprehensive study on microstructural, mechanical and corrosion properties. Mater Chem Phys. 2021;270:124839. https://doi.org/10.1016/j.matchemphys.2021.124839.
14. Migiakis K, Daniolos N, Papadimitriou GD. Plasma keyhole welding of UNS S32760 super duplex stainless steel: microstructure and mechanical properties. Mater Manuf Process. 2010;25(7):598–605. https://doi.org/10.1080/10426910903179955.
15. Westin EM, Johansson MM, Bylund LÅ, Pettersson RFA. Effect on microstructure and properties of super duplex stainless steel welds when using backing gas containing nitrogen and hydrogen. Weld World. 2014;58(3):347–54.
16. Bermejo MAV, Karlsson L, Svensson LE, Hurtig K, Rasmuson H, Frodigh M, Bengtsson P. Effect of shielding gas on welding performance and properties of duplex and superduplex stainless steel welds. Weld World. 2015;59:239–49.
17. Ferreira PM, Pereira EC, Pinheiro FW, Monteiro SN, Azevedo ARG. Effect of solution heat treatment by induction on UNS S31803 duplex stainless steel joints welded with the autogenous TIG process. Metal. 2022;12:1450.
18. Zhang ZH, Jing H, Xu L, Han Y, Zhao L, Lv X. Effect of post-weld heat treatment on microstructure evolution and pitting corrosion resistance of electron beam-welded duplex stainless steel. Corr Sci. 2018;141:30. https://doi.org/10.1016/j.corsci.2018.06.030.
19. Westin EM, Olsson COA, Hertzman S. Weld oxide formation on lean duplex stainless steel. Corr Sci. 2008;50:2620–34.
20. Ramkumar KD, Mishra D, Vignesh MK, Raj BG, Arivazhagan N, Naren SV, Kumar SS. Metallurgical and mechanical characterization of electron beam welded super-duplex stainless steel UNS 32750. J Manuf Process. 2014;16:527–34.
21. Tahaei A, Perez AFM, Merlin M, Valdes FAR, Garagnani GL. Effect of the addition of nickel powder and post weld heat treatment on the metallurgical and mechanical properties of the welded UNS S32304 duplex stainless Steel. Soldagem Inspeção. 2016;21(2):197–208. https://doi.org/10.1590/0104-9224/SI2102.09.

22. Rokanopoulou R, Skarvelis P, Papadimitriou GD. Welding design methodology for optimization of phase balance in duplex stainless steels during autogenous arc welding under Ar–N2 atmosphere. Weld World. 2018;63:3. https://doi.org/10.1007/s40194-018-0660-0.
23. Zhou Y, Mahmood S, Engelberg DL. Application of bipolar electrochemistry to assess the corrosion resistance of solution annealed lean duplex stainless steel. Mater Des. 2023;232: 112145. https://doi.org/10.1016/j.matdes.2023.112145.
24. Kim ST, Jang SH, Lee IS, Park YS. Effects of solution heat-treatment and nitrogen in shielding gas on the resistance to pitting corrosion of hyper duplex stainless steel welds. Corr Sci. 2011;53:1939–47.
25. Guttmann G. Intermediate temperature aging of duplex stainless steels. A review. Duplex Stainless Steels. 1991;1:79–92.
26. Hosseini VA, Thuvander M, Lindgren K, Oliver J, Folkeson N, Gonzalez D, Karlsson L. Fe and Cr phase separation in super and hyper duplex stainless steel plates and welds after very short aging times. Mater Des. 2021;210:110055.
27. Maurya AK, Pandey CH, Chhibber R. Dissimilar welding of duplex stainless steel with Ni alloys: a review. In J Press Vessel Pip. 2021;192:104439.
28. Hosseini VA, Wessman S, Hurtig K, Karlsson L. Nitrogen loss and effects on microstructure in multipass TIG welding of a super duplex stainless steel. Mater Des. 2016;98:88.
29. Hanninen H, Romu J, Ilila R, Tervo J, Laitinen A. Effects of processing and manufacturing of high nitrogen-containing stainless steels on their mechanical, corrosion and wear properties. J Mater Process Technol. 2001;117:424–30.
30. Abdollahzadeh A, Bagheri B, Abbasi M, Sharifi F, Mirsalehi SE, Moghaddam AO. A modified version of friction stir welding process of aluminum alloys: analyzing the thermal treatment and wear behavior. Proc Inst Mech Eng Part L J Mater Des Appl. 2021;235(10):2291–309. https://doi.org/10.1177/14644207211023987.
31. Abbasi M, Abdollahzadeh A, Bagheri B, Moghaddam AO, Sharifi F, Dadaei M. Study the effect of the welding environment on the dynamic recrystallization phenomenon and residual stresses during the FSW process of aluminum alloy. Proc Inst Mech Eng Part L J Mater Des Appl. 2021;235(8):1809–26. https://doi.org/10.1177/14644207211025113.
32. Abbasi M, Bagheri B, Abdollahzadeh A, Moghaddam AO. A different attempt to improve the formability of aluminum tailor welded blanks (TWB) produced by the FSW. Int J Mater Form. 2021;14(5):1189–208. https://doi.org/10.1007/s12289-021-01632-w.
33. Ogawa T, Koseki T. Effect of composition profiles on metallurgy and corrosion behavior of duplex stainless steel weld metals. Weld J. 1989;68:181–91.
34. Bagheri B, Abdollahzadeh A, Sharifi F, Abbasi M, Moghaddam AO. Recent development in friction stir processing of aluminum alloys: microstructure evolution, mechanical properties, wear and corrosion behaviors. Proc Inst Mech Eng Part E J Process Mech Eng. 2021; https://doi.org/10.1177/09544089211058007.
35. Rizi VS, Abbasi M, Bagheri B. Investigation into microstructure and mechanical behaviors of joints made by friction stir vibration brazing between low carbon steels. Phys Scrip. 2021;96 (12):125704. https://doi.org/10.1088/1402-4896/ac1dcb.
36. Bagheri B, Abbasi M, Sharifi F, Abdollahzadeh A. Investigation into novel multipass friction stir vibration brazing of carbon steels. Mater Manuf Process. 2022;37(8):921–32.
37. Graziano U, Firrao D, Taveri G. Characterization of welded joints (MIG and SAW) on LDX 2101 duplex SS. Process Eng. 2015;109:484–91.
38. Khan WK, Chhibber R. Effect of filler metal on solidification, microstructure and mechanical properties of dissimilar super duplex/pipeline steel GTA weld. Mater Sci Eng A. 2020;803: 140476. https://doi.org/10.1016/j.msea.2020.140476.

Chapter 7
Laser Welding of Lightweight Aerospace Alloys

Esad Kaya and Koray Kılıçay

7.1 Introduction

The term "laser" describes the physical process and propagation phenomenon of light radiation. It means the emission of stimulated radiation by the amplification of light. Lasers do not occur naturally [1]. It is light with electromagnetic waves and can only be produced artificially [2]. Scientists have accomplished this process at various wavelengths during the past 70 years. Gould [3] and Maiman [4] were the first to successfully emit at wavelengths of 694.3 and 692.9 nm, respectively. In addition, there are the theoretical and practical research projects of Einstein [5], Ladenburg [6], Schawlow and Townes [7], Gould [3], Javan [8], and numerous other researchers [1, 9].

Figure 7.1 depicts the general characteristics of lasers. Lasers can focus intensely on a small area. Their outstanding features include their high power density, excellent monochromaticity, the capacity to operate in limited frequency ranges, and the ability to focus on one focal point. Optical tools can focus a laser beam on a small area [9]. This situation leads to an indirect transformation of high-power density energy into a heat source. The laser energy makes it simple to heat, melt, or even evaporate ceramics, metals, composites, and plastics—some primary material categories. Laser processing can be used for hard, brittle materials that are challenging to shape using mechanical techniques, improving their shaping capabilities [1, 2, 9].

The idea of energy levels is a single factor influencing the laser's emission. Three primary mechanisms exist for the laser's interaction with the material's atoms, molecules, and ions [2]. These consist of stimulated emission, absorption, fluorescence, and spontaneous emission effects [2, 9]. Figure 7.2 shows the schematic

E. Kaya (✉) · K. Kılıçay
Department of Mechanical Engineering, Eskişehir Osmangazi University, Eskişehir, Turkey
e-mail: esatkaya@ogu.edu.tr; kkilicay@ogu.edutr

S. Gürgen (ed.), *Advances in Manufacturing for Aerospace Alloys*, Sustainable Aviation, https://doi.org/10.1007/978-3-031-64455-9_7

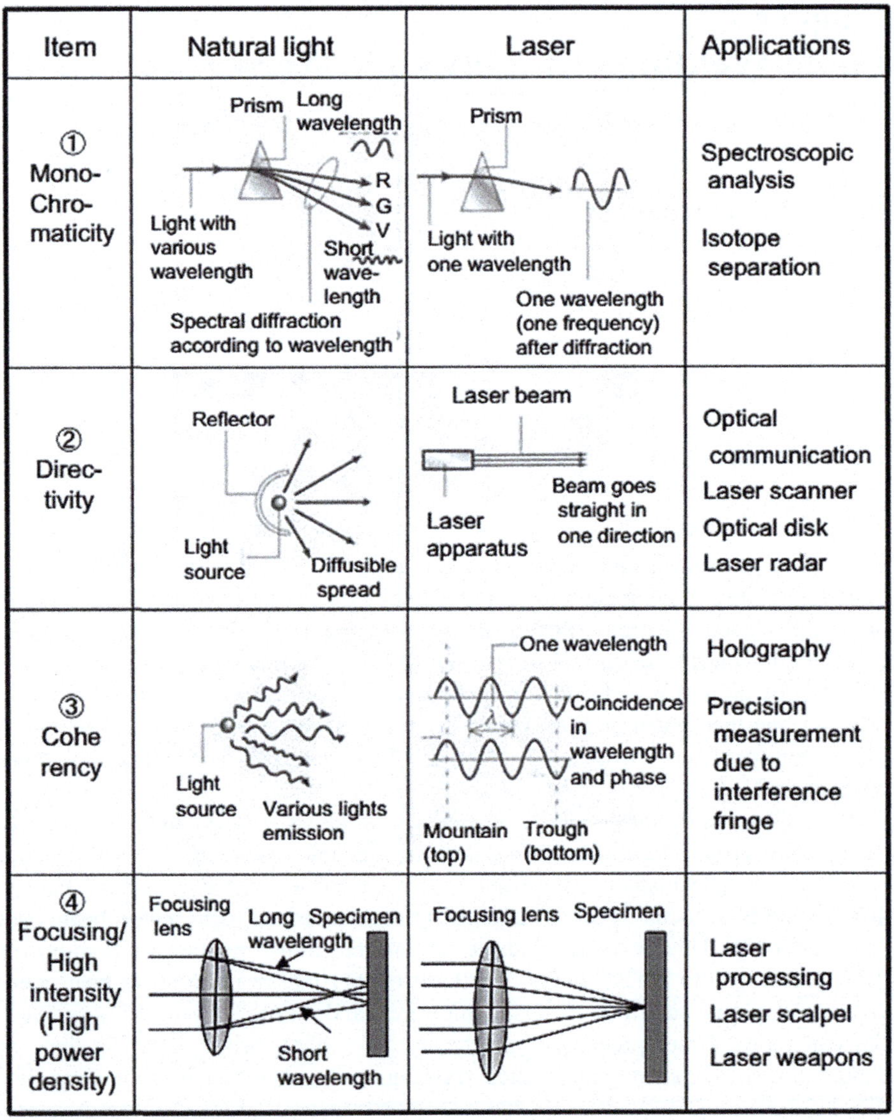

Fig. 7.1 Formation, physical features, and applications of laser beams [2]

illustration of these three scenarios. The low-energy atom absorbs the incoming light energy. The energy it absorbs raises it to a higher energy level. The governing equation, Eq. 7.1, controls the event's physics.

$$E = h\nu \tag{7.1}$$

where E, h, and ν represent the light's energy, constant of Planck's, and frequency, respectively [1, 2, 9].

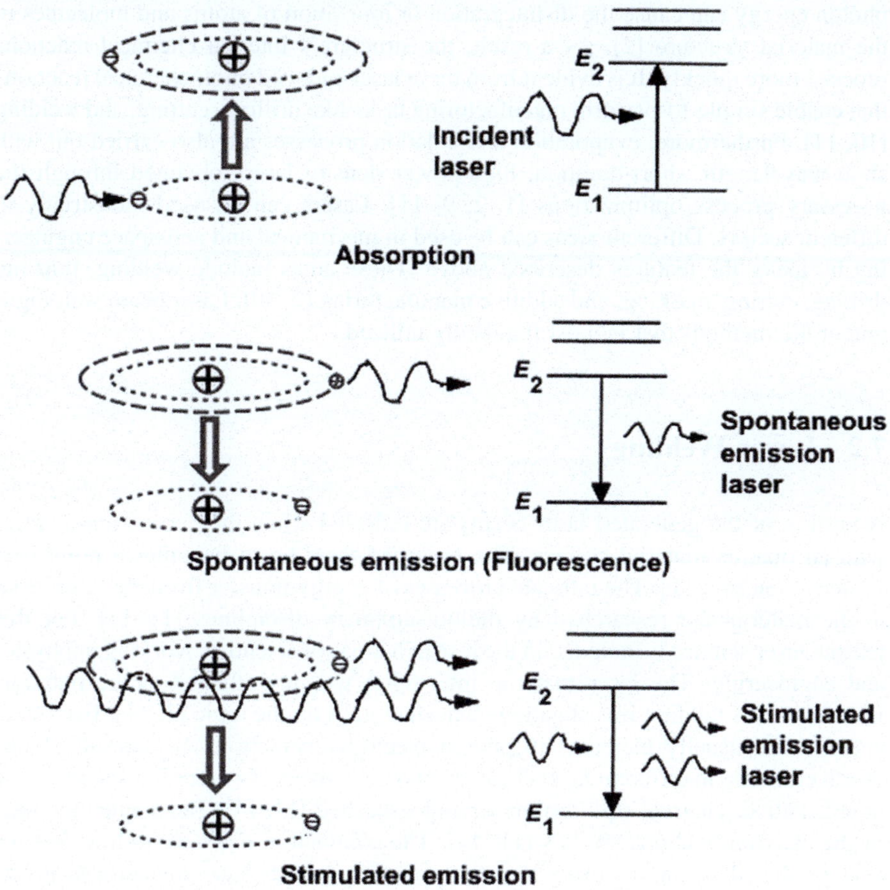

Fig. 7.2 Three main mechanisms for laser interactions [2]

The atom is present at the highest energy level and decays spontaneously under fluorescence or spontaneous emission. It drops to a more minimal energy state. The photon with the right frequency disperses the energy difference between them. In the case of stimulated emission, the incoming light interacts with the atom at the upper energy level. It ensures that light is released at a lower energy level and a frequency equivalent to the energy differential. In the relevant electric field, the frequency, phase, and direction of motion of these atoms are the same. As the light intensity increases, laser emission originates from this process [1, 2, 9].

Laser beams pass through vacuums and other atmospheric conditions quickly. Processing various materials and sizes without physical contact across long distances becomes possible by the laser process [2, 9]. With laser technology and optimally determined parameters, operations can be performed on the inner surfaces of polymers, glass ceramics, and similar transparent materials [2]. The processes are not affected by the magnetism of the material. Lasers with short wavelengths but high photon energy can cause the disintegration or ionization of atoms and molecules in the material structure [2]. As a result, the structure's internal chemical reactions proceed more quickly. It is evident from these laser-induced photochemical reactions that enable simple-to-perform manufacturing tasks like drilling, cutting, and welding [10, 11]. Furthermore, evaporation and ablation processes are also carried out with short-wavelength, short-duration, high-power density lasers obtained through the necessary process optimizations [1, 2, 9, 11]. Lasers can be used industrially in different sectors. Different areas can be used in mechanical and aerospace engineering by using the features described above. These areas include welding, brazing, drilling, cutting, marking, and additive manufacturing [2, 10]. Laser beam welding is one of the methods that is most frequently utilized.

7.2 Laser Welding

A portion of the generated laser beams are reflected when they come into contact with an opaque material surface. The absorbed laser beam becomes a point heat source as energy [11]. The intricate processes of energy transfer from the laser beam to the material are researched by multidisciplinary disciplines [11–13]. For this reason, laser science is included in the research areas of scientists working in physics and engineering. The laser–material interaction was investigated at the photonic level in one of the first investigations that shed light on the topic [2, 11]. Electrons' energy state usually increases in photon–electron interaction. As a result of this event, electrons in different layers heat up, and then, the energy rapidly spreads to the atomic lattice. Thermal equilibrium is rapidly reached [11, 14]. The conduction time of the electron is about 10^{-13} s [11, 14]. The conduction mechanism transfers the heat produced by laser beams that the radiation mechanism concentrates to the surface. This case leads to the occurrence of local heating. Localized melting continues, and a phase transition develops from enhanced sustained constant heat transfer [2]. The materials are fused. In some methods, electronic, mechanical, or hybrid technologies may all be utilized to control the continuation of heat transfer. It is acknowledged that indirect manipulations in any heat transfer can be used for cutting, welding, or surface-alloying [11]. Figure 7.3 schematically represents the process of laser welding. The first studies focus on laser welding with deep penetration dating back to the 1960s, performed by Sickman and Meriju [15]. Arata and Miyamoto [16] investigated the behavior of molten regions following intense penetration. Studies have found that heat transfer occurs in a narrow, thin, and long way in the liquefied areas due to the intense penetration of the laser source. When sufficient melting power is achieved on the solid surfaces contacted by the laser, a hole is observed from the surface of the material inward. The pressure generated by the resulting melting and even evaporation causes the molten metal to migrate along the hole wall [11, 13]. Melting proceeds, and the cavity is maintained because of the constant laser emission. The gap that forms at the locations where the laser beam meets the solid surface causes a continuous material

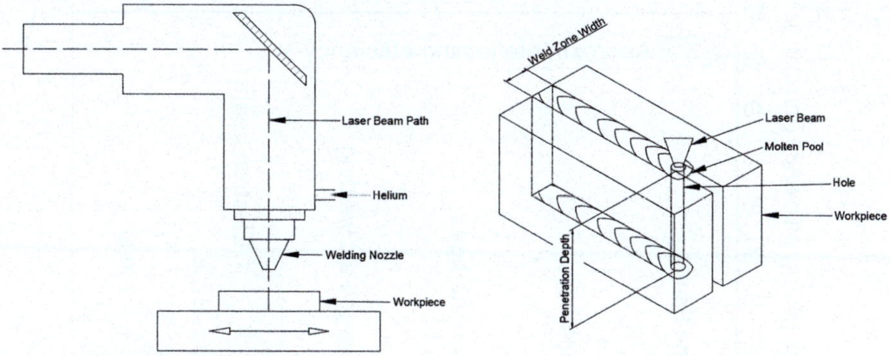

Fig. 7.3 Schematic illustration of laser welding on material and weld bead formation

fluidity. The metal that was pouring out of the gap freezes as a result of the moving laser source, and a wall surrounds the hole [11, 13]. The beam source advances in the created molten void region at a predetermined speed when it is moved. Only with steady-speed movement of the continuously melting and solidifying zone can penetration be made that will result in a successful binding. In order to maintain the resulting hole in a stable state, adequate evaporation must be guaranteed [1, 11, 13]. The maximum feed rate for which this steady state can be achieved must be determined. Welding speed can be determined numerically and experimentally by considering the combination of the material to be welded and power density. For this reason, it is advised that the welding engineers and operators establish on the parameters, including penetration depth and speed, for the material that has to be welded [1, 2, 11].

One of the most efficient welding methods with the highest power density utilized in industrial applications is focused laser beam welding. Its high energy density is an application point of new technology, even if its power density is similar to electron beam welding [13]. With laser welding, evaporation takes place due to the high energy density on the material's surface. Evaporation proceeds while the welding operation is ongoing and subsequently ends with cooling. The laser effect produces keyhole from the surface to deep inside the material [2, 11]. Welding transition zones, which are known as fusion zones, are almost parallel to each other. Fusion zones are characterized by their narrowness and depth compared with the other welding methods [2, 11, 17, 18].

Figure 7.4 shows the joining efficiency values for various welding techniques. As may be observed, laser and electron beam welding techniques yield the highest joining efficiency ratings. A comparison of the welding types reveals that laser welding has the region under the narrowest heat effect. Melting the interfaces that need to be linked is crucial when assessing the effectiveness of the welding [13, 17]. Joining efficiency is the term for this definition of efficiency used to assess the efficacy of various welding process types. Joining efficiency is expressed as a unit area joined per energy transferred [13]. Although it is not an actual efficiency

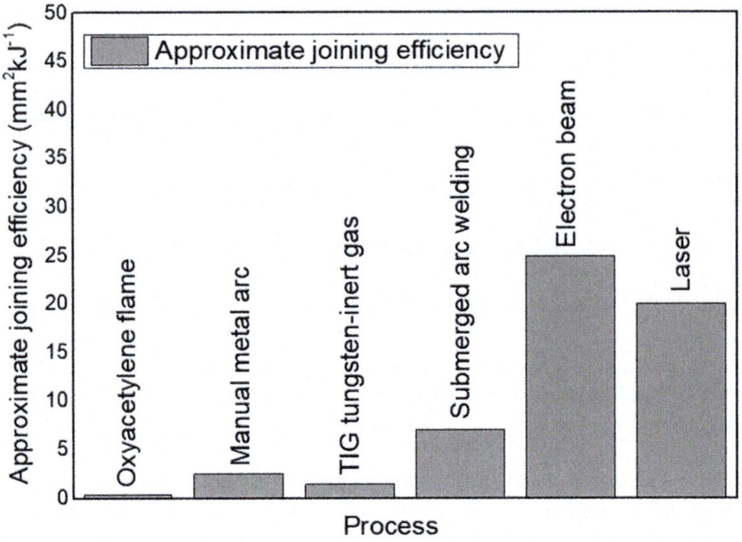

Fig. 7.4 Joining efficiency values provided for different welding methods [13]

Welding Process	Flux-shielded arc		Gas-shielded arc		Plasma		Laser or electron beam	
Heat source intensity (w/m^2)	$5x10^6$	$5x10^8$	$5x10^6$	$5x10^8$	$5x10^6$	$5x10^{10}$	10^{10}	10^{12}
Fusion zone profile			low	high	low	high	defocus	focus

Fig. 7.5 Heat source intensity and fusion zone profile of different welding methods [13]

term, it provides a numerical comparison. Joining efficiency is dependent on welding speed (V), weld thickness (t), and incident power (P) as given in Eq. 7.2.

$$\text{Joining Efficiency} = V.t/P \tag{7.2}$$

Transferring energy to the areas other than the joining surfaces causes heating of these areas, heat-induced microstructure changes, and indirect problems such as distortion, internal stress, microcracks, and low strength. These areas are called heat-affected zones (HAZ). It is understood that the higher the joining efficiency in a welding method, the less the amount of energy transferred to the HAZ. When Fig. 7.5 is considered, it is seen that the highest energy input is in the laser welding method along with the electron beam welding. It is also seen that the areas with the smallest HAZ are in the laser welding method. This method has a very narrow HAZ, especially compared with the other gas welding methods [2, 13].

Table 7.1 Characteristics and advantages of laser welding [13]

Characteristics	Advantages
High energy density—"keyhole" type weld	Less distortion
High processing speed	Cost-effective (if fully employed)
Rapid start/stop	Unlike arc process
Welds at atmospheric pressure	Unlike electron beam welding
No X-rays generated	Unlike electron beam
No filler required (autogenous weld)	No flux cleaning
Narrow weld	Less distortion
Relatively little HAZ	Can weld near heat-sensitive materials
Possibility of very accurate welding	Can weld thin to thick materials
Good weld bead profile	No clean-up necessary
No beam wander in the magnetic field	Unlike electron beam
Little or no contamination	It depends only on gas shrouding.
Relatively little evaporation loss	Advantages with Mg and Li alloys
Feasible for complex weld materials	General superiority
Suitable for automatization	General superiority
Time shared system	General superiority

Table 7.1 lists the characteristics and advantages of laser welding. To begin with, the HAZ geometry of laser welding differs from that of all other welding techniques. Its structure resembles a keyhole, which characterized it. High production speeds in manufacturing are made possible by skilled workers. It is simple to achieve instant stops or continuous output. Unlike electron beam welding, it can be done at atmospheric pressure and radiation-free. In essence, low HAZ formation is ensured by powerful coupling efficiency. It works particularly well with light metals that are challenging to weld. Compared with other welding processes, excellent weld seams with shallow contamination are obtained under the protective gas whose ideal conditions are established [2, 13, 18].

It is evident from an extensive analysis that laser welding offers numerous benefits. Mass production has specific applications these days. Using the characteristics mentioned above, lasers are used in various settings. It is extensively utilized in additive manufacturing, coating, and welding processes [2].

7.2.1 Laser Sources and Types

Researchers and laser manufacturers are constantly improving industrial lasers. Lasers are named according to the environment in which they will be used: propagation medium, material properties, oscillation medium, wavelength, and pulse frequency. Lasers are classified as gas, liquid, and solid. Solid lasers are also known as Nd:YAG (neodymium-doped yttrium aluminum garnet; Nd:$Y_3Al_5O_{12}$) and are ideal for ease of use. Transmission can be achieved up to 1–200 m with different lasers. Wavelength varies depending on the type [2, 17].

Lasers with higher power were first developed using CO_2. Subsequently, excimer-type XeCl (308 nm), KrF (238 nm), and ArF (193 nm) lasers were developed. Plastic, glass, and similar materials are the main materials that these lasers are used for drilling and cutting. In the following years, high-power diode lasers with wavelengths of 0.8–1.1 μm were developed. These types of lasers used for material processing have been called direct diode lasers (DDLs) or high-power direct diode lasers (HPDDLs). In addition to these, disc and fiber lasers have also been developed. High-quality and high-power diode lasers are used as the pump source in these types of lasers. In the last two decades, disk- and fiber-type lasers with high power and efficiency have been developed. These lasers are widely used, especially in welding and cutting processes. In addition, the produced laser is used continuously depending on time and by fluctuating the laser power. This use is called the pulse method. In recent studies, pulse welding processes are at the picosecond and femtosecond levels. While ultra-short pulse lasers are used for drilling operations, they are also used for lap welding of glasses [2, 17]. Table 7.2 lists the characteristics of the fundamental laser types.

Table 7.2 The characteristics of lasers for welding [2]

	Wavelength (μm)	Laser media	Average power (kW)	Merit
CO_2 laser	10.6, far-infrared ray	CO_2–N_2–He mixture	0.5–12	More accessible high power, 10–20% efficiency
Lamp-pumped YAG laser	1.06, near-infrared ray	Nd^{3+}:$Y_3Al_5O_{12}$ (solid)	0.05–5	Fiber delivery and easier handling, good flexibility, 1–4% efficiency
Laser Diode (LD)	1.0, near-infrared ray	Al(In)GaAs, (solid)	2–60	Compact, fiber delivery, and high 20–50% efficiency
LD-pumped solid-state laser	1.06, near-infrared ray	Nd^{3+}:$Y_3Al_5O_{12}$ (solid)	6–13.5	Fiber delivery, high brightness, and high 10–20% efficiency
Disk laser	1.03, near-infrared ray	Yb^{3+}:YAG or YV04 (solid)	16	Fiber delivery, four branches, high brightness, 15–25% efficiency
Fiber laser	1.07, near-infrared ray	Yb^{3+}:SiO_2 (solid)	10–120	Compact, fiber delivery, high brightness, 20–50% efficiency
Green laser	515, 532, 535 nm	Nonlinear optical crystal from disk, YAG, or fiber laser (solid)	0.5–3	Short wavelength, high absorption for copper, 5–20% efficiency
Ultrashort pulse laser	0.8–1.6, near-infrared ray	Titanium–sapphire, fiber laser, and disk laser (solid)	0.1–1	Ultrashort pulse, extremely narrow HAZ

7.2.2 CO_2 Laser

One of the most used industrial lasers is the CO_2 type, which has a wavelength of 10.6 μm and higher output power. It also has a higher efficiency than Nd:YAG lasers. It has excellent beam quality and is simple to focus. However, complex and reliable manipulation methods are needed due to light. These lasers cannot be transmitted via fiber optics. Despite the weak absorption of the wavelength, it has high output powers (up to 50 kW) enough to create keyholes in high-thickness metals. CO_2, N_2, He, and CO gases are generally used to create a CO_2 laser. This quaternary mixture is discharged as incandescent under 0.1 atm or lower pressures. Figure 7.6 shows the energy levels for CO_2 laser emission. CO_2 laser with a wavelength of 10.6 μm has a triatomic bond structure. It spreads through asymmetric stretching vibrations and transitions from upper to lower energy levels. N_2 gas contained in the quaternary mixture contributes to the rise in the number of CO_2 molecules in the upper energy levels [2, 17].

In addition, laser power is stabilized by glow discharge. Power losses and discharge imbalances that may occur in the CO_2 laser are eliminated with CO gas. Since water and moisture reduce the power of the laser, they must be removed from the environment. Apart from this, by choosing the reflectivity of the mirrors, the CO_2

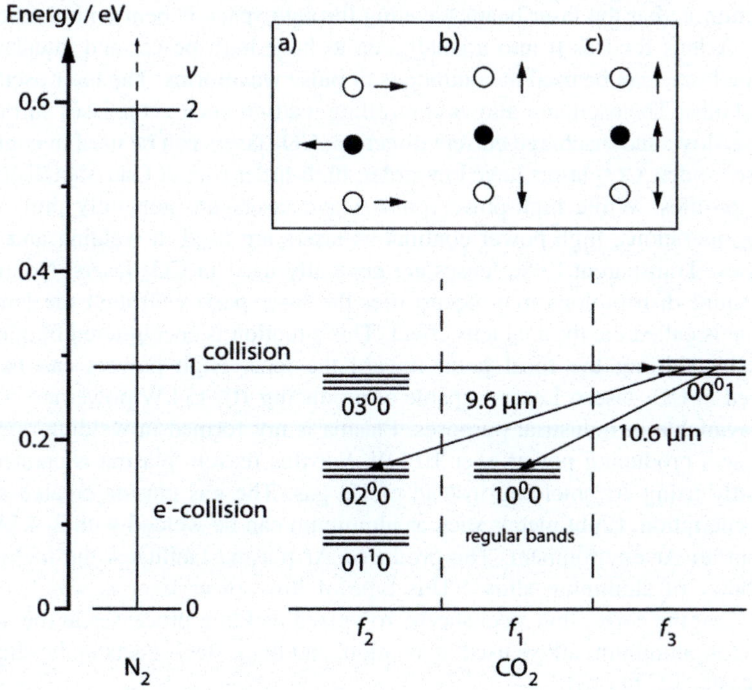

Fig. 7.6 Energy levels for CO_2 laser [19]

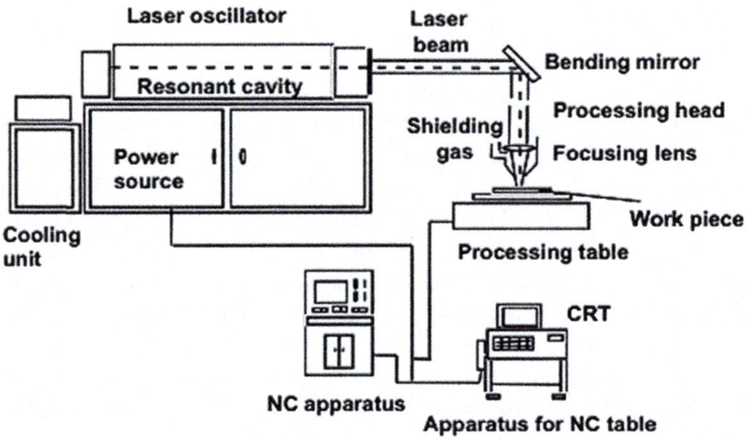

Fig. 7.7 Schematic illustration of the CO_2 laser welding equipment [2]

laser can also be emitted at a wavelength of 9.6 μm due to bending vibration. The laser at this wavelength is mainly used in pulse conditions, drilling, and similar operations on printed circuit boards [2, 17, 19].

Figure 7.7 shows the schematic of the CO_2 laser device. The basic logic in laser production is that the laser beam traveling through space is bent with the help of a mirror. A lens focuses it into a small area to have high power and intensity. The resulting lasers can be used in continuous or pulse waveforms. The laser oscillator is a sealed tube. The oscillator allows controlling features such as the laser's irradiation axis, gas flow, and discharge current direction. CO_2 lasers can be used in continuous or pulse modes. CO_2 lasers have low-order, high-order ring or Gaussian distribution power profiles. While high-power pulse wave lasers are generally preferred for drilling operations, high-power continuous lasers are used in welding and cutting processes. Transparent ZnSe lenses are generally used in CO_2 lasers. Nonuniform temperature distributions may occur after the laser passes through the lens. This situation is called the thermal lens effect. This situation is encountered frequently. It occurs by shifting the focal point toward the lens. High powers can be easily achieved in CO_2 lasers. Lasers capable of producing 10–45 kW power are commercially available in industrial purposes. Plasma is not formed in welding operations with lasers producing power over 10 kW. For this reason, plasma is created more efficiently using Ar shielding instead of He gas. The gas plasma created shallow weld penetration. Light metals such as aluminum can be welded with 2–4 kW CO_2 lasers in an Ar environment. The produced Ar plasma facilitates the melting and weldability of aluminum alloys. This type of low-power laser is widely used in pulleys, engine cores, thin steel sheets, robotized welding processes in the automotive sector, aluminum alloys used in aviation, and thick steel sheets in the shipbuilding industry [2, 11, 13].

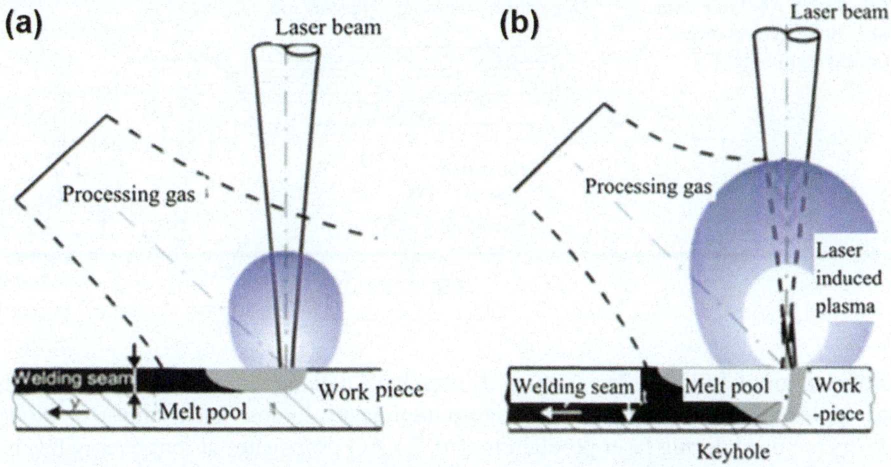

Fig. 7.8 The modes of the laser beam welding process: (**a**) conduction and (**b**) keyhole [22]

Joining in laser welding can be completed in two ways. One of them is the conduction mode, which is shown in Fig. 7.8a. As the wavelength of the laser decreases, the absorbency of the metallic material increases. CO_2 lasers have absorption problems because they contain higher wavelengths than Nd:YAG lasers. Therefore, the conduction mode is often used to manufacture dimensionally small parts. The produced laser beam is dropped on the workpiece with a small number of optical apparatus. The beam's wavelength in CO_2 lasers requires optical devices instead of fiber optic cables. By allowing a gap to form in the welding area, that is, when the keyhole is created, the projected diameter is focused on. Figure 7.8b shows the keyhole mode. There is molten material in conical form around the resulting keyhole. This resulting gap is in balance with the material's ablation and the molten material's hydrostatic pressure. What is required to maintain this balance is the movement speed of the laser and, indirectly, the source. Hence, the minimum and maximum speeds required for balance must be determined. Welding applications performed at speed higher than the maximum speed cause collapse around the keyhole, while insufficient speed causes the areas under the influence of heat to grow excessively. As can be seen, the keyhole concept is deeply dependent on many factors. For this reason, fluctuations in shape and size may occur during welding, which may detrimentally affect the desired strength factors [2, 20, 21].

7.2.3 YAG Laser

The YAG laser takes its name from the ion form of the rare earth elements Nd^{+3} and Er^{+3} and the yttrium aluminum garnet (YAG; $Y_3Al_5O_{12}$) rod contribution. This form of laser is generally of the Nd^{+3}:YAG type. It is typically an Nd^{+3}:YAG laser with a

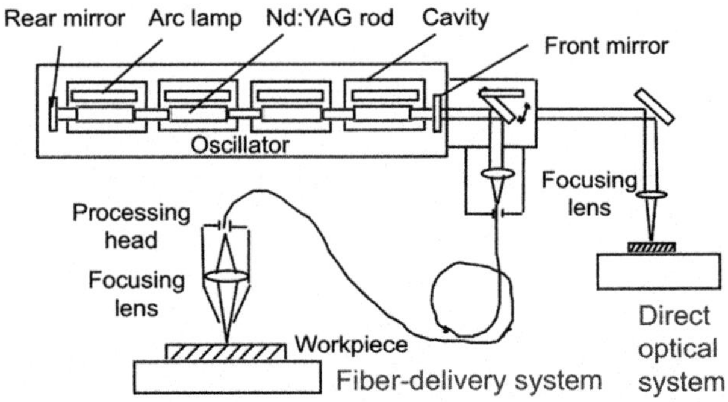

Fig. 7.9 YAG laser with fiber delivery and direct optical equipment [2]

wavelength of 1.06 µm. The YAG rods used have a diameter of 10 mm and a length of approximately 100–150 mm. Light-pumping arc lamps, flash lamps, or diode lasers are used during laser production. Nd^{+3}:YAG propagates at four energy levels. Yttrium Vanadate (YVO_4) laser and Yttrium Lithium Fluoride (YLF) laser types are similar. Figure 7.9 shows a schematic illustration of the working principle and apparatus of a YAG laser. Direct optical systems or transmission methods via fiber optics are used for welding. Lasers with a wavelength of 1.06 µm are used in material processing processes. Graded index-type laser has a low damage threshold. They have lower power than 500 W but provide better quality beam transmission. Step index-type lasers are used for different processes and applications requiring higher power [2, 17].

7.2.4 Diode Laser

Diode lasers are semiconductors with a double heterojunction structure. There is a current from the outside to the semiconductor on the inside. Laser is obtained by emitting electrons from the active layer in the n-type and p-type semiconductor regions. Figure 7.10 schematically shows the working principle of the diode laser. Photons are moved according to the degree of refraction of the mirrors used. Laser is achieved through stimulated emission due to amplification. This type of laser is also called semiconductor or diode laser in different sources. The advantages of diode laser are as follows [2].

- Compact and lightweight system due to their small volume.
- Laser efficiency exceeding 50%.
- Long lifetime.
- Power combinations due to different current techniques.
- Low cost due to suitability for mass production.

The disadvantages of diode lasers are as follows.

- Low laser power from a single device.
- Inability to focus.

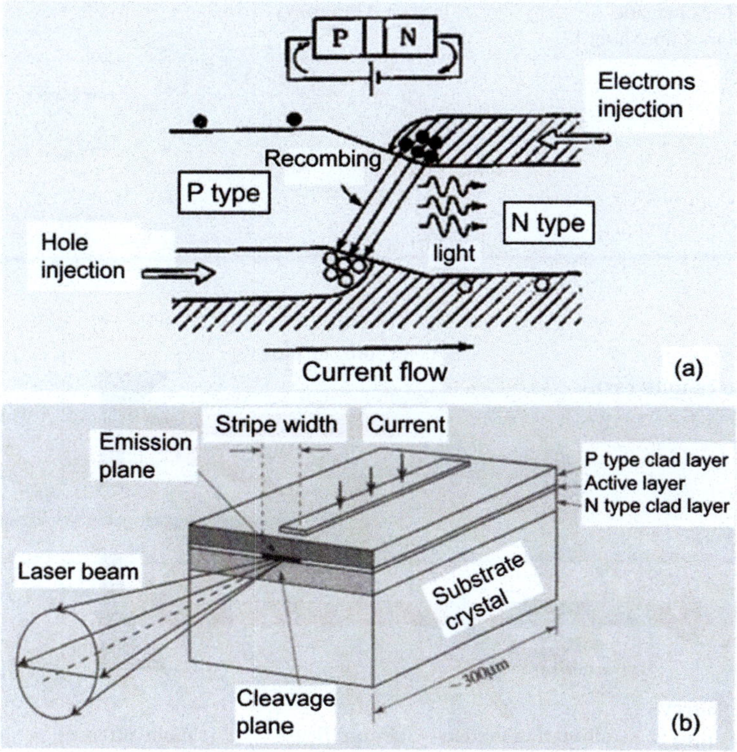

Fig. 7.10 Diode laser applications on the material surfaces [2]

According to the most recent investigations and studies, the drawbacks have mainly been addressed. Since higher power lasers have been developed, splitting and dichroic mirrors have been employed for high power production, merging fibers and laser beams, employing semiconductor rods, and polarizing beams of different wavelengths. Diode lasers are employed in laser cladding and martensitic transformation of steels to connect thin sheets. It has also recently been used in deep penetration welding processes on aluminum alloys. [2].

7.2.5 *Fiber Laser*

Fiber lasers were initially put into production in the 1960s. The first 100-W fiber laser was produced in 2000. Today, types of lasers used for material processing up to 30 kW limits have been produced. The production of higher power systems is ongoing due to technological advancements. High-energy power fiber lasers have numerous benefits. These consist of mobilization, high beam quality with tiny beam

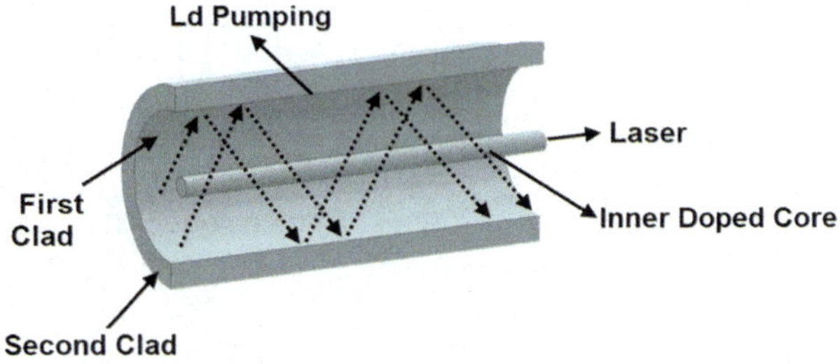

Fig. 7.11 Schematic illustration of fiber laser [2]

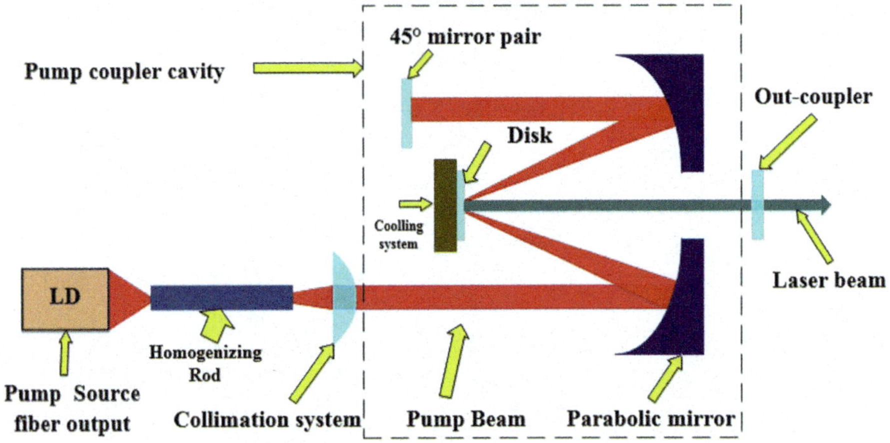

Fig. 7.12 Schematic illustration of disk-type laser [24]

diameters, compact designs, and high efficiency. Figure 7.11 depicts the fiber laser's operating principle. The active medium is the core containing rare earth elements. Silica fiber contains Er, Yb, Nd, or Tl elements. Multimode diodes are used in the pumping process. Two Bragg layers on the fiber core limit the beam wavelength. The resulting laser beam has low diffraction and produces a highly parallel beam [2, 17, 23].

7.2.6 Disk Laser

Disk lasers are the state-of-the-art solid-state lasers. Figure 7.12 schematically depicts the operational mechanism. The central apparatus used in disk lasers is the YAG crystal, which has a thickness of 150–300 μm and a diameter of 12–15 mm. This crystal is usually Yb-doped. The heat generated during laser beam generation is dispersed on the Cu radiator. The laser disk is mounted on this Cu radiator. The YAG disk is pinpoint-focused in an area of 1–2 mm^2. The diode is illuminated and optically excited by laser radiation. In practice, diode lasers match the pumped

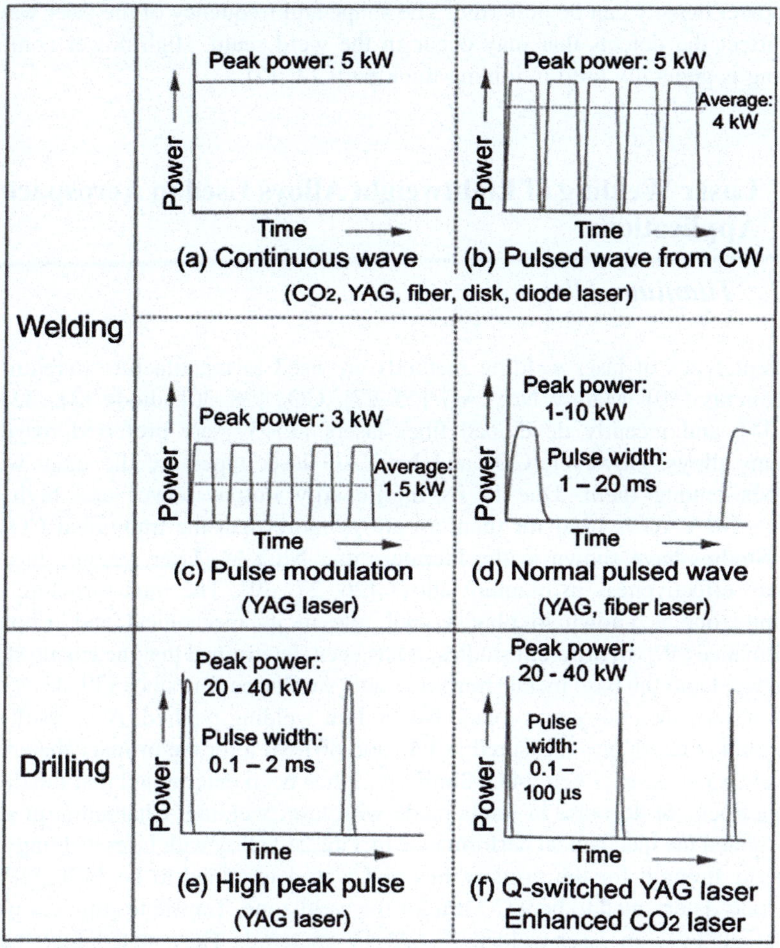

Fig. 7.13 Various characteristic waves of continuous and pulsed welding for operations [2]

radiation to the wavelength. Since the disk used is very thin, some of the radiation passing through the disk is absorbed. The part that cannot pass is reflected by the disk and directed back onto the disk with the installed mirror optical system [2, 17].

The many types of laser sources listed can generate beams in various ways. The laser beam can also be sent to the material surface in a continuous or pulsed form, as shown in Fig. 7.13. Both conduction type and keyhole form can be formed, depending on the achieved laser power density. The weld seams are shallow when the laser power intensity is low. They range in depth from 0.6 to 2 mm. In cases where laser power density is high, keyhole morphology occurs. The keyhole format varies between approximately 0.5–3 mm and 0.5–50 mm for both pulsed keyhole types and continuous applications. Therefore, pulse welding can be used for materials with low thickness and slow speeds, or continuous welding with high speed and low power density can be preferred. The shape and frequency of the laser wave can also affect the defects that may occur in the weld seam. High-power continuous welding is generally used in joining thick parts [2, 17].

7.3 Laser Welding of Lightweight Alloys Used in Aerospace Applications

7.3.1 Titanium Alloys

Different types of laser welding methods are used to manufacture titanium alloy components [20]. Mainly Nd:YAG [25–27], CO_2 [28–30], diode [31, 32], disk [33, 34], and recently developed fiber lasers [35–37] are preferred for joining titanium alloys. However, CO_2 and Nd:YAG laser types are the main welding methods among them. Due to its high-energy output power and high beam quality, Nd:YAG is used for titanium alloys more than the traditional CO_2 laser [38]. Studies have shown in the literature that Nd:YAG laser welding has better welding effectiveness in titanium alloys [26, 27, 39]. The microstructure of the welding zone is known to play a vital role in its mechanical and tribological performance [40]. In different studies, it has been determined that the tensile strength of the weld and the base metal showed almost similar performance [39, 41–43]. In a study, $Ti_2Al_{1.5}Mn$ alloy was joined by the butt welding method. As a result of the tests, joint strength was measured as 642 and 661 MPa for the primary material and the weld zone, respectively [44]. Similarly, it has been determined that the ductility change does not decrease in joints made with laser welding. Chamanfar et al. [45] investigated the mechanical performance of Ti62242 alloy with laser welding. While the yield strength for the primary material was determined to be 1120 MPa, this value was determined to be 977 MPa for the weld zone. Tensile testing was used by Cao et al. [39] on laser-welded Ti-6Al-4V material. They stated that the tests revealed comparable performance in the tensile and yield strengths. Nonetheless, they discovered that the ductility had nearly doubled, and there was a noticeable embrittlement in the weld area. According to Kumar et al. [36], the ultimate tensile strength and elongation for Ti-6Al-4V welds were 872 MPa and 3.74%, respectively. These values were less than the base metal's respective values of 957 MPa and 11%. The experiments clarify that poor parameter selections and weld structural flaws deteriorate mechanical performance. Laser-welded titanium alloys exhibit reductions, particularly in ductility performance. Based on a compilation of studies, it is believed that oxide formations and microporosity caused by aluminum in the titanium alloy structure are the primary causes of the weld area's increased brittleness [20].

Ti-6Al-4V is one of the most used metallic alloys in the aerospace industry [20]. In laser welding, welding speed and power directly affect the penetration depth. This subject is widely researched in the literature due to the different elements

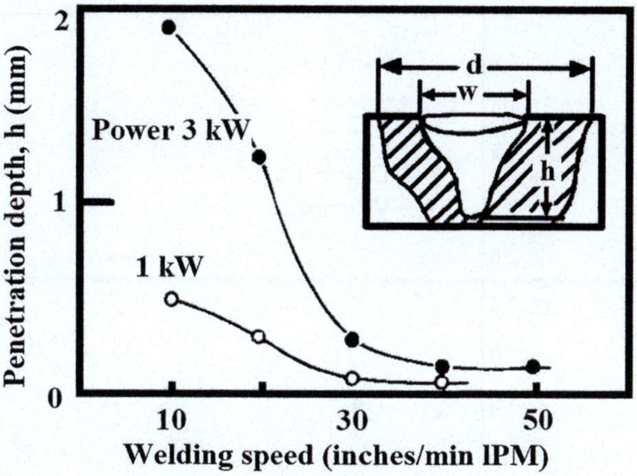

Fig. 7.14 The effect of welding speed and laser energy input on the penetration depth [11]

contained in titanium alloys [20]. Figure 7.14 depicts the impact of welding power and speed on penetration depth. Penetration tends to decrease with increased welding speed while power is constant. As expected, the penetration depth rose as power increased at a constant welding pace [20]. In a study led by Akman et al. [27], they investigated how a pulsed Nd:YAG laser affected the weld quality of Ti-6Al-4V alloy. According to the study, the wavy form of the laser beam provided more profound penetration depth results than the continuous form. Another fiber laser study demonstrated significant welding speed limitations. They discovered that the welding shape is in V-form in fiber laser sources with a power of 1.2 kW between 2 and 2.5 m/min [46]. They found that increasing welding speed increased the risk of closed porosity formation. In other studies, they found that using a high-power fiber laser for Ti-6Al-4V alloy caused undercut and spatter defects. Similarly, they found that high welding speed caused incomplete penetration [47, 48]. The Ti-6Al-4V alloy's microstructural characteristics after CO_2 laser welding were investigated. In the studies, CO_2 laser and gas metal (such as tungsten inert gas [TIG] and plasma) welding methods were compared based on weld morphology and microstructural features. Weld seam morphology is directly correlated with speed. As expected, the smallest heat-affected zone grain sizes were seen in laser-welded specimens [20, 49].

7.3.2 Aluminum Alloys

Aluminum alloys have extensive applications in the aerospace industry [21]. In today's industrial settings, the nonferrous material category is most frequently used. Compared with iron alloys, particularly steels, it has a lower melting point and density. Furthermore, according to Punkari et al. [50], it possesses higher thermal expansion, resilient surface oxide, and strong corrosion resistance [51]. Compared with steel, it also boasts a high strength-to-weight ratio

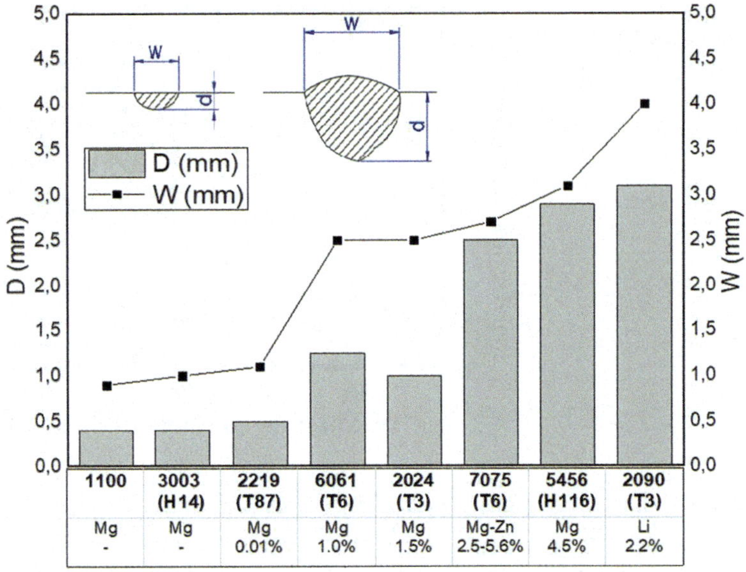

Fig. 7.15 The comparison of penetration depth and welding beads for different aluminum alloys after CO_2 laser welding [54]

[52]. Furthermore, compared with other metallic materials, it performs exceptionally well in terms of cost-effectiveness and recyclability [53]. Due to their superior performance, they are used primarily in aviation, automobiles, ships, and places where the speed–light ratio is crucial [21]. There are two different methods of producing aluminum alloys for industrial use. These methods are rolling and casting methods [21]. In addition, aluminum alloys are divided into two groups based on their suitability for heat treatment. While the types unsuitable for heat treatment can gain strength by rolling and deformation hardening, the strength of the other part can be increased by aging processes. The heat conductivity of aluminum alloys is around four times that of iron alloys. Its ability to deflect lasers is likewise higher at the same time. These circumstances make it challenging to use the welding process for aluminum alloys, particularly the laser welding method [21]. For this reason, parameter determination processes such as penetration depth, welding speed, and power of aluminum alloys are far more crucial. Figure 7.15 shows the penetration depths after laser welding and the widths of the areas affected by the welding on the surface for aluminum alloys with different chemical compositions [2, 21].

Structural efficiency in components used in aviation is vital in terms of fuel economy and payload-carrying capacity [55]. Different methodologies are being tried to reduce flying weight. These are mainly effective mechanism designs and the design of new-generation materials. In particular, material-based improvements are achieved using lower density new materials [55]. Aluminum alloys are widely used in aviation due to their low density–strength ratio. Today, new-generation Al–Li containing alloys are special aluminum alloys developed for aviation [56]. These alloys are characterized by their high strength and particularly low density compared with conventional aluminum alloys. Due to the mentioned features, it has been developed for the last 50 years, and different companies and scientists have produced new generations. They can be manufactured by hot, cold, and superplastic forming techniques after the casting and homogenization process. In addition, Weldalite® and similar third-generation alloys can also be processed by laser welding methods [21, 55, 57]. Although the forging process yields parts with better mechanical performance than welding and casting, it could not be cost-effective enough for parts with intricate geometry [57, 58].

7.3.3 Dissimilar Welding of Titanium and Aluminum Alloys

Despite the differences in melting temperatures, titanium and aluminum alloys can be joined by welding [20]. It has been determined through different studies that such dissimilar alloys can be welded using the keyhole method and laser welding method [59–62]. Using keyholes in dissimilar welding processes provides flexibility and controllability to the welding process. In addition, the metallurgical reaction time is reduced as high welding speed is used in keyhole-form processes [63]. The formation of brittle intermetallic material in the region where it melts and solidifies quickly and in a short time is significantly reduced with laser welding. However, the laser beam offset in butt joints is known as the main parameter that controls the thickness of the intermetallics to be formed. In the recent studies, many researchers have suggested shifting the laser beam to the titanium side. As it is known, the melting temperature of titanium alloys is higher. This way, the titanium alloy side is treated with keyhole mode, while the aluminum alloy side is treated with conduction mode [60, 61, 64]. Kreimeyer et al. [64] joined 1.15-mm-thick AA6016T4 and 0.8-mm Ti-6Al-4V alloys together by using a commercially pure titanium (CP-Ti; Grade 2) wire using CO_2 laser welding. They characterized the microstructural and mechanical properties of the weld. The study found that shifting the laser beam 0.3 mm to the titanium side minimized the melting amount of the aluminum alloy. In this way, they found that the static strength increased by 200 MPa. Using optimum parameters, they reduced the formation thickness of intermetallic layers to levels below 2 μm as shown in Fig. 7.16.

Elements such as Mg, Zn, and Li found in aluminum alloys increase the alloy's ability to melt and penetrate more easily. As a result of the increased penetration effect, easier melting and wider weld seams can be achieved [54]. A profound keyhole creates a deeper weld seam. Figure 7.17 shows the weld zone microstructures of AA5083 alloy joined by laser welding at different welding speeds and power. In the laser welding process, the risk of porosity decreases at low speeds, and the weld expands. At high speeds, most bubbles produced at the tip of the hole melted and escaped to the pool surface [2, 65].

The welding zone microstructure is also extremely influenced by the welding circumstances. Any drastic parameter alteration impacts the microstructure and

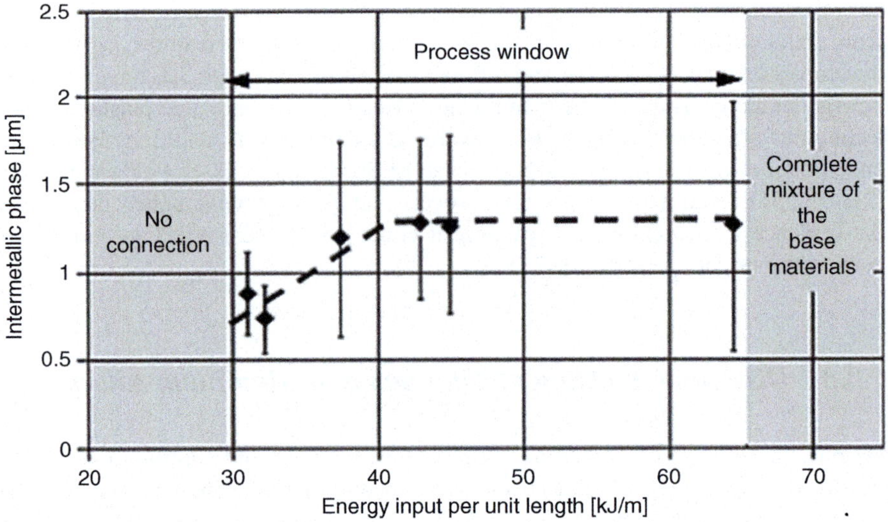

Fig. 7.16 The optimum process gap for the laser-welded AA6016T4 and Ti-6Al-4V for tailored blanks [64]

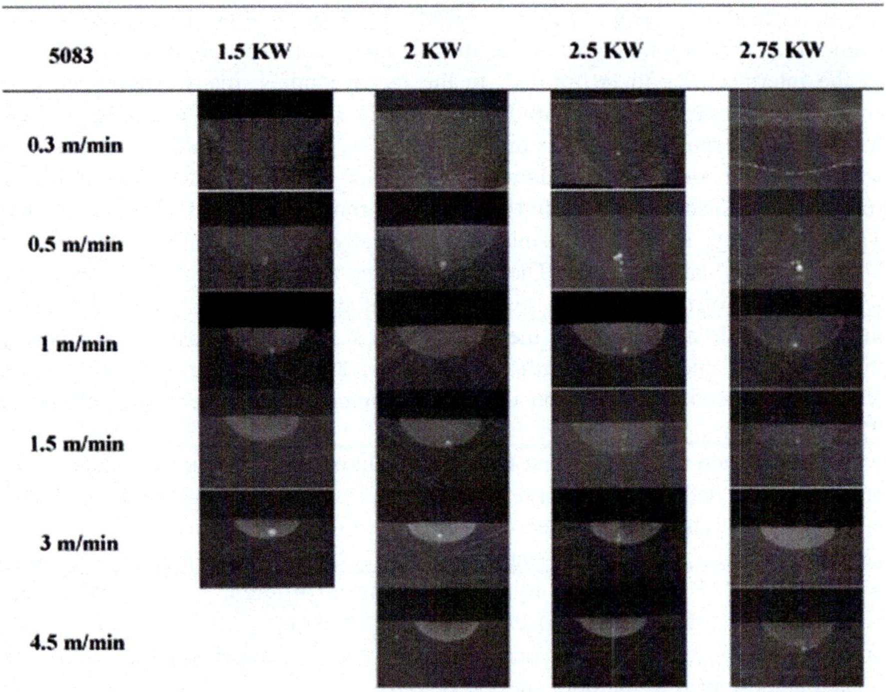

Fig. 7.17 The effect of laser energy input and speed on the fusion zone of AA5083 alloy [65]

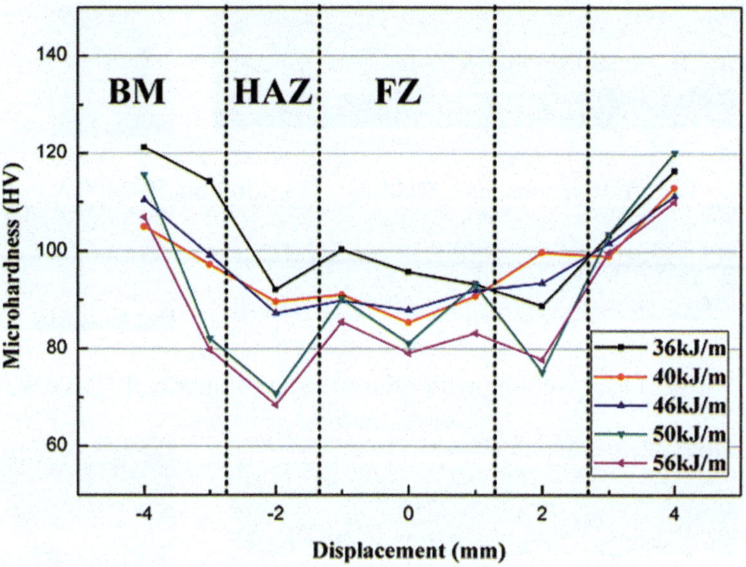

Fig. 7.18 The microhardness profile of the laser-welded AA2060, 2 mm thickness and 3 m/min at different energy inputs [66]

similarly affects the mechanical properties of the welding zone and the heat-affected zones [54]. Figure 7.18 shows the hardness profile for an Al–Li alloy, AA2060 with laser welding at different powers. As can be seen, the highest hardness values were measured in the base metal, while the lowest hardness was recorded in the region under the influence of heat. In general, hardness decreases in the central weld zone due to the alloying elements in the structure. The decrease in the area under the influence of heat is the grain coarsening caused by thermal distortion and the resulting softening. Although increasing heat input in the welding process provides surface bonding, it also causes lower microhardness distribution [66].

Good macro- and micro-form weld joints can be produced with process parameters that are well designed and determined through experimentation. Figure 7.19 shows laser welding applications of Al–Li alloys manufactured with ideal specifications.

If the welding process parameters are not determined optimally, some classical defect types can be seen as macro and micro on the weld seams. The most common defects seen as macro are excessive penetration, undercut, and surface porosity. Figure 7.20 shows typical welding defects encountered in laser welding. When the distance between the focal point and the wire becomes zero or negative, discontinuities occur in the weld (Fig. 7.20a). Incorrect selection of power causes pilling in the weld (Fig. 7.20b). If filler metal is used and the tip of the wire blocks the laser focus, the energy cannot melt the main material, and large droplets are formed (Fig. 7.20c). The macro-surface pores formed are mainly caused by partial wire feeding. If filler metal falls into the back end bath of the laser, wire stay occurs

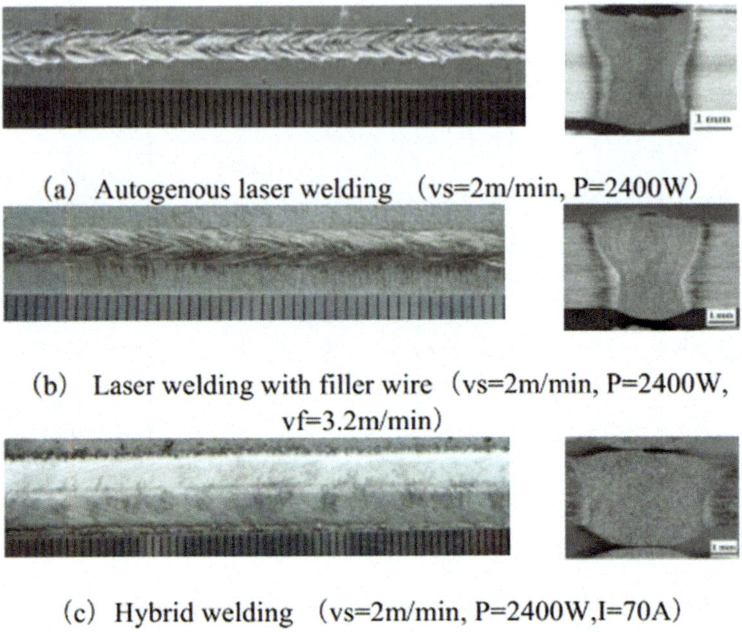

Fig. 7.19 Laser-welded joints using different techniques [67]

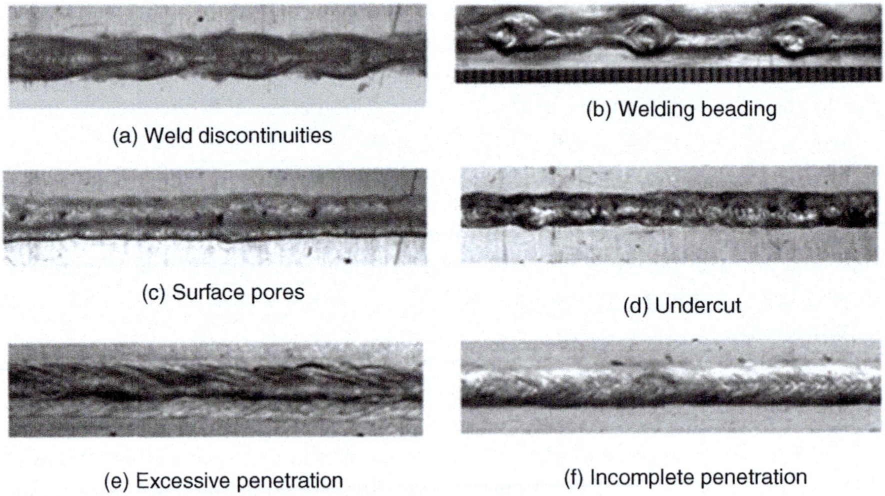

Fig. 7.20 Typical defects in laser-welded surfaces [67]

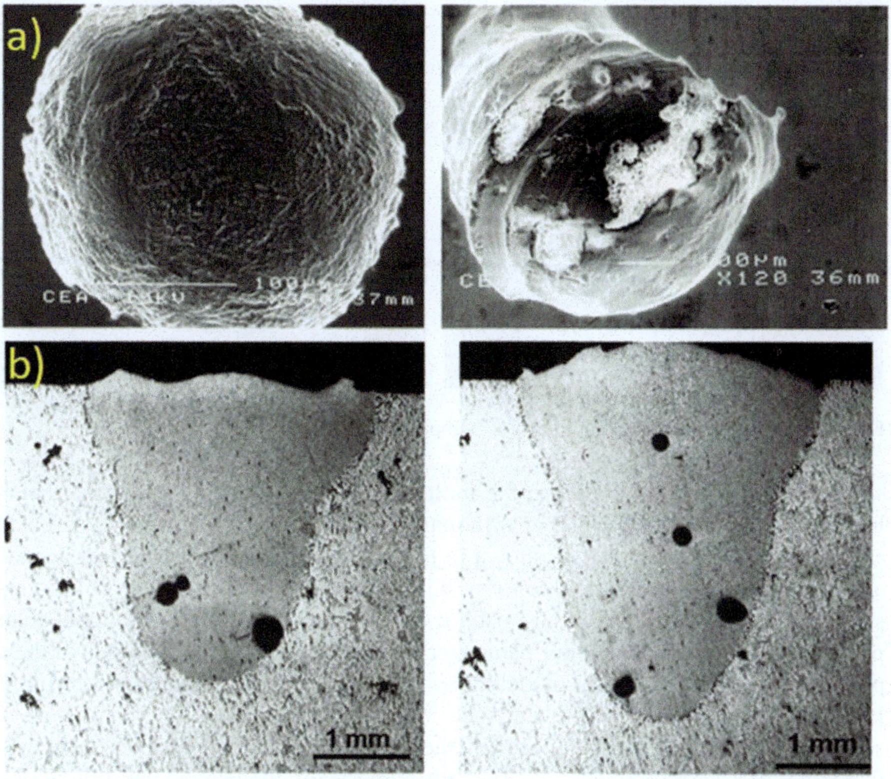

Fig. 7.21 (**a**) Porosities caused by inclusions formed after laser welding and (**b**) closed void structures formed due to incorrect parameter selection [68]

(Fig. 7.20d). With too high energy input and a low filler metal feed rate, deposits form (Fig. 7.20e). In cases where the energy is deficient, defects and incomplete penetration occur (Fig. 7.20f). These described defects are prevented by optimizing the parameters in the process [67].

Produced weld seams should be controlled from both macro and micro views. In general, alloying elements within the structure can form different compounds with the environment, causing closed porosities. While these are Li_2O, LiOH, Li_2CO_3, and LiN compounds for Al–Li alloys, it can be said that they are MgO compounds for magnesium alloys. Figure 7.21 shows the porosities caused by inclusions after laser welding and the void structures caused by incorrect parameter selection [21, 67, 68].

In weld microstructures, porosity problems primarily fall into two categories. The first one is smaller than the other. The porosity of the crystallization layer is this kind of imperfection, showing a typical example in Fig. 7.21a. The other kind of inaccuracy is the vast gaps that arise from the lack of time for hydrogen and other

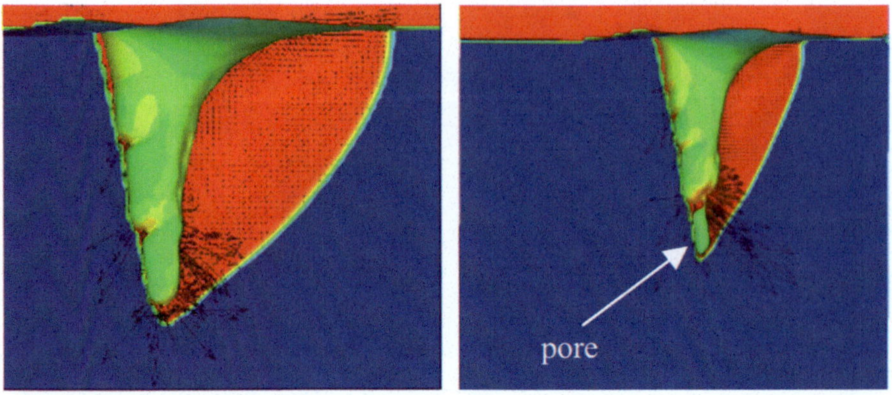

Fig. 7.22 Numerical simulation of keyhole and pore formations in laser welding [67]

small atomic diameters to escape the molten pool. These defects are of the metallurgical variety. Nucleation and growth mark the start of the formation stages. These are typical closed oval areas, as shown in Fig. 7.21b. The laser beam causes a small and deep penetration weld pool to form. It is trapped between the rapidly solidifying strata during the pool's flow and cannot escape the surface [21, 68]. Inadequate preparation procedures and poor parameter selection are the primary causes of these problems. Various numerical modeling and simulations are typically utilized in the laser welding procedure to prevent such inaccuracies. An example of this kind of scenario is shown in Fig. 7.22. Based on the principles of fluid mechanics, the region that liquefies as a result of energy transfer is numerically modeled in the simulations. It simulates the behavior of the metal at a specific temperature and pressure when it is molten. According to the simulations, these kinds of pores are typically created by the movement of the keyhole during the welding process. It happens due to the keyhole encircling the molten metal vapor and gas mixture. Process parameter incompatibilities give rise to these pores [67, 69].

There are various techniques for identifying these kinds of porosity structures. X-ray methods and numerical analyses can be used to detect it. The level of defects in the structures of Al–Li alloys welded with various parameters is shown in Fig. 7.23. Metallurgical-type flaws unavoidably arise in the welding area if no surface treatment is performed before laser welding. The array shape exhibits significantly more gaps than other shapes, as demonstrated in Fig. 7.23a, b. The inadequate surface preparation before welding is the primary cause of these gaps' enormous size. Figure 7.23c shows gaps that are similar to these. On the other hand, the frequency and form of voids are declining. Insufficient penetration is the cause of situations like this one. The ideal conditions shown in Fig. 7.23d–f can be observed if sufficient safety measures are performed before welding and full penetration are accomplished [67, 68].

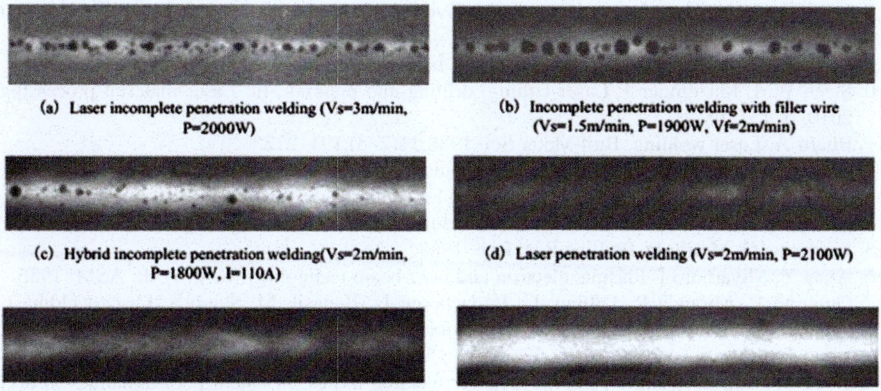

(a) Laser incomplete penetration welding (Vs=3m/min, P=2000W)

(b) Incomplete penetration welding with filler wire (Vs=1.5m/min, P=1900W, Vf=2m/min)

(c) Hybrid incomplete penetration welding(Vs=2m/min, P=1800W, I=110A)

(d) Laser penetration welding (Vs=2m/min, P=2100W)

(e) Penetration welding with filler wire (Vs=2m/min, P=2400W, Vf=2m/min)

(f) Hybrid penetration welding (Vs=2m/min, P=1800W, I=130A)

Fig. 7.23 X-ray inspection photos of laser-welded joints at different parameters [67]

7.4 Conclusions

This chapter investigated laser sources and laser formation, discussing general principles for laser beam welding of lightweight aerospace alloys. By considering microstructural and mechanical performances, the benefits of laser welding are described by avoiding the challenges brought about by traditional methods. The effects of the laser welding methods on different lightweight aerospace materials, such as aluminum and titanium alloys, are discussed. In addition to discussing joints of similar alloys, this chapter also investigates laser welding joints involving dissimilar alloys. Identifying potential welding defects is explained using examples from the relevant literature.

References

1. Blackburn J. Laser welding of metals for aerospace and other applications. In: Welding and joining of aerospace materials; 2012. p. 75–108.
2. Katayama S. Kinds and characteristics of lasers for welding. Fundamentals and details of laser welding. In: Topics in mining, metallurgy and materials engineering; 2020. p. 1–17.
3. Gould RG. The LASER, light amplification by stimulated emission of radiation. In: The Ann Arbor conference on optical pumping; 1959.
4. Maiman TH. Stimulated optical radiation in ruby. Nature. 1960;187(4736):493–4.
5. Einstein A. Strahlungs-emission und -absorption nach der quantentheorie. Verhandl D Deutch Phys Ges. 1916;18:318–23.
6. Ladenburg RW. Anomalous dispersion in excited gases. I. Proof of the dispersion formula in excited neon. Z Phys. 1928;48:15–25.
7. Schawlow AL, Townes CH. Infrared and optical masers. Phys Rev. 1958;112(6):1940–9.

8. Javan A. Possibility of production of negative temperature in gas discharges. Phys Rev Lett. 1959;3(2):87–9.
9. Steen WM, Mazumder J. Basic laser optics. In: Laser material processing; 2010. p. 79–130.
10. Steen WM, Mazumder J. Laser cutting, drilling and piercing. In: Laser material processing; 2010. p. 131–98.
11. Bharti A. Laser welding. Bull Mater Sci. 1988;11(2–3):191–212.
12. Anderson JEJJ. Theory and application of pulsed laser. Weld J. 1965;44(12):1018–26.
13. Steen WM, Mazumder J. Laser welding. In: Laser material processing; 2010. p. 199–249.
14. Ready JF. Effects of high-power laser radiation 1971.
15. Sickman JG, Meriju R. Phillips Res. Rep. 1968.
16. Arata Y, Miyarnoto I. Plasma, electron and laser beam technology. New York: ASM; 1986.
17. Quintino L, Miranda R, Dilthey U, Iordachescu D, Banasik M, Stano S. Laser welding of structural aluminium. In: Structural connections for lightweight metallic structures. Advanced structured materials; 2010. p. 33–57.
18. Gong S, Pang S, Wang H, Zhang L. Laser welding basics. In: Weld pool dynamics in deep penetration laser welding; 2021. p. 1–18.
19. Eichler HJ, Eichler J, Lux O. Infrared molecular gas lasers. In: Lasers: basics, advances and applications; 2018. p. 89–110.
20. Auwal ST, Ramesh S, Yusof F, Manladan SM. A review on laser beam welding of titanium alloys. Int J Adv Manuf Technol. 2018;97(1–4):1071–98.
21. Oladimeji OO, Taban E. Trend and innovations in laser beam welding of wrought aluminum alloys. Weld World. 2016;60(3):415–57.
22. Kumar C, Das M, Biswas P. A 3-D finite element analysis of transient temperature profile of laser welded Ti-6Al-4V alloy. In: Lasers based manufacturing. Topics in mining, metallurgy and materials engineering; 2015. p. 421–40.
23. Miyamoto I, Park S-J, Ooie T. Ultrafine-keyhole welding process using single-mode fiber laser. In: International congress on applications of lasers & electro-optics. AIP Publishing; 2003.
24. Liu R, Zhang X, Gong F, Jia Y, Li G. Research on the adjusting technology of the thin disk laser. Optik. 2018;157:400–5.
25. Gao X-L, Zhang L-J, Liu J, Zhang J-X. Porosity and microstructure in pulsed Nd:YAG laser welded Ti6Al4V sheet. J Mater Process Technol. 2014;214(7):1316–25.
26. Gao X-L, Zhang L-J, Liu J, Zhang J-X. A comparative study of pulsed Nd:YAG laser welding and TIG welding of thin Ti6Al4V titanium alloy plate. Mater Sci Eng A. 2013;559:14–21.
27. Akman E, Demir A, Canel T, Sınmazçelik T. Laser welding of Ti6Al4V titanium alloys. J Mater Process Technol. 2009;209(8):3705–13.
28. Shi Y, Zhong F, Li X, Gong S, Chen L. Effect of laser beam welding on fracture toughness of a Ti-6.5Al-2Zr-1Mo-1V alloy sheet. J Mater Sci. 2007;42(16):6651–7.
29. Balasubramanian TS, Balasubramanian V, Muthu Manickam MA. Fatigue crack growth behaviour of gas tungsten arc, electron beam and laser beam welded Ti–6Al–4V alloy. Mater Des. 2011;32(8–9):4509–20.
30. Caiazzo F, Curcio F, Daurelio G, Minutolo FM. Ti6Al4V sheets lap and butt joints carried out by CO2 laser: mechanical and morphological characterization. J Mater Process Technol. 2004;149(1–3):546–52.
31. Amaya-Vazquez MR, Sánchez-Amaya JM, Boukha Z, Botana FJ. Microstructure, microhardness and corrosion resistance of remelted TiG2 and Ti6Al4V by a high power diode laser. Corros Sci. 2012;56:36–48.
32. Lisiecki A. Welding of titanium alloy by different types of lasers. Arch Mater Sci Eng. 2012;58 (2):209–18.
33. Caiazzo F, Alfieri V, Fierro I, Sergi V. Investigation and optimization of disk-laser welding of 1 mm thick Ti-6Al-4V titanium alloy sheets. Adv Mech Eng. 2014;7(2)
34. Caiazzo F, Cardaropoli F, Alfieri V, Sergi V, Argenio P, Barbieri G. Disk-laser welding of Ti-6Al-4V titanium alloy plates in T-joint configuration. Procedia Eng. 2017;183:219–26.

35. Costa A, Miranda R, Quintino L, Yapp D. Analysis of beam material interaction in welding of titanium with fiber lasers. Mater Manuf Process. 2007;22(7–8):798–803.
36. Kumar C, Das M, Paul CP, Singh B. Experimental investigation and metallographic characterization of fiber laser beam welding of Ti-6Al-4V alloy using response surface method. Opt Lasers Eng. 2017;95:52–68.
37. Nakai M, Niinomi M, Akahori T, Hayashi K, Itsumi Y, Murakami S, et al. Microstructural factors determining mechanical properties of laser-welded Ti–4.5Al–2.5Cr–1.2Fe–0.1C alloy for use in next-generation aircraft. Mater Sci Eng A. 2012;550:55–65.
38. Cao X, Jahazi M, Immarigeon JP, Wallace W. A review of laser welding techniques for magnesium alloys. J Mater Process Technol. 2006;171(2):188–204.
39. Cao X, Jahazi M. Effect of welding speed on butt joint quality of Ti–6Al–4V alloy welded using a high-power Nd:YAG laser. Opt Lasers Eng. 2009;47(11):1231–41.
40. Lütjering G. Influence of processing on microstructure and mechanical properties of (α+β) titanium alloys. Mater Sci Eng A. 1998;243(1–2):32–45.
41. Wang SH, Wei MD, Tsay LW. Tensile properties of LBW welds in Ti–6Al–4V alloy at evaluated temperatures below 450 °C. Mater Lett. 2003;57(12):1815–23.
42. Liu H, Nakata K, Yamamoto N, Liao J. Mechanical properties and strengthening mechanisms in laser beam welds of pure titanium. Sci Technol Weld Join. 2011;16(7):581–5.
43. Liu J, Watanabe I, Yoshida K, Atsuta M. Joint strength of laser-welded titanium. Dent Mater. 2002;18(2):143–8.
44. Fang X, Liu H, Zhang J. Microstructure and mechanical properties of pulsed laser beam welded Ti-2Al-1.5Mn titanium alloy joints. J Mater Eng Perform. 2014;23(6):1973–80.
45. Chamanfar A, Pasang T, Ventura A, Misiolek WZ. Mechanical properties and microstructure of laser welded Ti–6Al–2Sn–4Zr–2Mo (Ti6242) titanium alloy. Mater Sci Eng A. 2016;663:213–24.
46. Campanelli SL, Casalino G, Mortello M, Angelastro A, Ludovico AD. Microstructural characteristics and mechanical properties of Ti6Al4V alloy fiber laser welds. Procedia CIRP. 2015;33:428–33.
47. Mueller SB, Mueller P, Cuddy J, Shankar K. Laser beam welding of titanium—a comparison of CO2 and fiber laser for potential aerospace applications. In: Proceedings of ICALEO. 27th international congress on applications of lasers and electro-optics; 2008. p. 846–54.
48. Ahn J, Chen L, Davies CM, Dear JP. Parametric optimisation and microstructural analysis on high power Yb-fibre laser welding of Ti–6Al–4V. Opt Lasers Eng. 2016;86:156–71.
49. Sun ZP, Zhang W. Correlation between welding parameters and microstructures in TIG, plasma and laser welded Ti-6 Al-4 V alloy. In: 6th international conference: trends in welding research; 2002. p. 760–7.
50. Punkari A, Weckman DC, Kerr HW. Effects of magnesium content on dual beam Nd : YAG laser welding of Al—Mg alloys. Sci Technol Weld Join. 2003;8(4):269–81.
51. Vargel C. Physical properties of aluminium. In: Corrosion of aluminium; 2004. p. 19–20.
52. Ramasamy S, Albright CE. CO2 and Nd–YAG laser beam welding of 5754–O aluminium alloy for automotive applications. Sci Technol Weld Join. 2001;6(3):182–90.
53. Lee CH, Kim SW, Yoon EP. Electron beam welding characteristics of high strength aluminium alloys for express train applications. Sci Technol Weld Join. 2000;5(5):277–83.
54. Katayama S. Features of laser welding or joining of various materials. In: Fundamentals and details of laser welding. Topics in mining, metallurgy and materials engineering; 2020. p. 145–65.
55. Prasad NE, Gokhale AA, Wanhill RJH. Aluminum-lithium alloys processing, properties, and applications. Butterworth-Heinemann: Elsevier; 2014. p. 608.
56. Olga Grushko BO, Ovchinnokov V. Brief history of aluminum–lithium alloy creation. In: Aluminum-lithium alloys. CRC Press; 2016. p. 1–10.
57. Kılıç S, Kacar İ, Öztürk F. Havacılık Endüstrisinde Yeni Trend: Al-Li Alaşımları. Gazi Üniversitesi Mühendislik-Mimarlık Fakültesi Dergisi. 2018;34(1):275–96.

58. Dorin T, Vahid A, Lamb J. Aluminium lithium alloys. In: Fundamentals of aluminium metallurgy. Woodhead Publishing; 2018.
59. Tomashchuk I, Grevey D, Sallamand P. Dissimilar laser welding of AISI 316L stainless steel to Ti6–Al4–6V alloy via pure vanadium interlayer. Mater Sci Eng A. 2015;622:37–45.
60. Majumdar B, Galun R, Weisheit A, Mordike BL. Formation of a crack-free joint between Ti alloy and Al alloy by using a high-power CO2 laser. J Mater Sci. 1997;32(23):6191–200.
61. Casalino G, Mortello M, Peyre P. Yb–YAG laser offset welding of AA5754 and T40 butt joint. J Mater Process Technol. 2015;223:139–49.
62. Tomashchuk I, Sallamand P, Cicala E, Peyre P, Grevey D. Direct keyhole laser welding of aluminum alloy AA5754 to titanium alloy Ti6Al4V. J Mater Process Technol. 2015;217:96–104.
63. Chen S, Yang D, Li M, Zhang Y, Huang J, Yang J, et al. Laser penetration welding of an overlap titanium-on-aluminum configuration. Int J Adv Manuf Technol. 2016;87(9–12): 3069–79.
64. Kreimeyer M, Wagner F, Vollertsen F. Laser processing of aluminum–titanium-tailored blanks. Opt Lasers Eng. 2005;43(9):1021–35.
65. Sánchez-Amaya JM, Delgado T, González-Rovira L, Botana FJ. Laser welding of aluminium alloys 5083 and 6082 under conduction regime. Appl Surf Sci. 2009;255(23):9512–21.
66. Gu C, Wei Y, Zhan X, Zhang D, Ren S, Liu H, et al. Investigation of welding parameters on microstructure and mechanical properties of laser beam-welded joint of 2060 Al–Cu–Li alloy. Int J Adv Manuf Technol. 2016;91(1–4):771–80.
67. Yang J, Gang S, Li X, Chen L, Xu F. Typical joint defects in laser welding of aluminium-lithium alloy. In: Hinduja S, Li L, editors. Proceedings of the 36th international MATADOR conference. London: Springer London; 2010.
68. Haboudou A, Peyre P, Vannes AB, Peix G. Reduction of porosity content generated during Nd: YAG laser welding of A356 and AA5083 aluminium alloys. Mater Sci Eng A. 2003;363(1–2): 40–52.
69. Gong S, Pang S, Wang H, Zhang L. Model of Quasi-steady weld pool dynamics and numerical simulation. In: Weld pool dynamics in deep penetration laser welding; 2021. p. 19–64.

Chapter 8
Direct Metal Laser Sintering of Aeroengine Materials

Abdul Hasib Hasan Zayed, Mostofa Jawad Itmum, Niaz Mohammad Zahin, Mahatab Bin Rashid, and Md Enamul Hoque

8.1 Introduction

As the efficiency of the aircraft depends much on the engine, the engine has always been the focus. Although conventional design methods such as forging, machining, and casting are successful in manufacturing a lot of components, when it comes to the manufacturing of complex parts such as the components of the jet engine, achieving precise design can be either too costly or simply impossible. This is where additive manufacturing comes as a practical solution to this problem. Aerospace applications have long made use of additive manufacturing. The first industrial rapid prototyping machinery was created by three-dimensional (3D) systems in the late 1980s. Among the first five clients was the American aerospace contractor Pratt & Whitney [1]. In contrast to traditional subtractive methods of manufacturing, additive manufacturing is a totally new method that has had a significant impact on product design, manufacture, and assembly in the aviation industry. Over the past few decades, the use of additive manufacturing has expanded across a wide range of industries. Using this additive manufacturing process, it is easy-to-fabricate complex parts. The process of computer numerical control (CNC) machining and additive manufacturing are comparable as well. Both the processes use computer-generated input data, but in the case of additive manufacturing, metal is added to the built sections, whereas in CNC machining the material is subtracted. In contrast to

A. H. H. Zayed · M. J. Itmum · N. M. Zahin
Department of Mechanical Engineering, Military Institute of Science and Technology (MIST), Dhaka, Bangladesh

M. B. Rashid
Department of Mechanical Engineering, University of Texas at Arlington, Arlington, TX, USA

M. E. Hoque (✉)
Department of Biomedical Engineering, Military Institute of Science and Technology (MIST), Dhaka, Bangladesh

S. Gürgen (ed.), *Advances in Manufacturing for Aerospace Alloys*, Sustainable Aviation, https://doi.org/10.1007/978-3-031-64455-9_8

traditional casting, additive manufacturing offers superior dimensional precision, operational flexibility, resource utilization, and the efficiency of material. Furthermore, it is a layer-by-layer manufacturing process [2–5].

One of the additive manufacturing techniques is the direct metal laser sintering (DMLS). It is an excellent substitute for machining or casting when producing parts with intricate designs quickly. It gives a designer the design flexibility of 3D printing along with the mechanical qualities associated with standard aerospace materials. Using a 3D model as a basis, a high-energy laser melts metal powder layer by layer to create one or more designs. The parts are created in a closed build room filled with nitrogen or argon gas [6]. It sinters the product's cross-section using ytterbium fiber laser light and gas-atomized metal powder [7]. The advantages of the DMLS process over conventional manufacturing processes are that the DMLS process can manufacture complex parts in a simple way by eliminating the tooling operations. It also takes less time, and the precision is high [6]. As it is an additive method, there is no waste of material. The flexibility of tool-less production combined with the geometric intricacy they offer is attractive; nevertheless, with complicated overhanging geometric shapes, complete geometrical flexibility is a goal but not achievable [8]. The advantages of DMLS were demonstrated in smaller scale, internally cooled turbine rotors and heat exchangers for sixth-generation fighter engines. Heat exchangers are constructed with thin walls and layers that are constrained by brazing and welding assembly procedures, which may result in residual tensile stresses that undermine the material's integrity. Likewise, casting process limits on cooling wall thickness dictate the size of integrally bladed, small-scale turbine rotors [9].

DMLS is a process for creating objects using powder-based materials that are difficult to form from the 3D computer models. Figure 8.1 shows a schematic

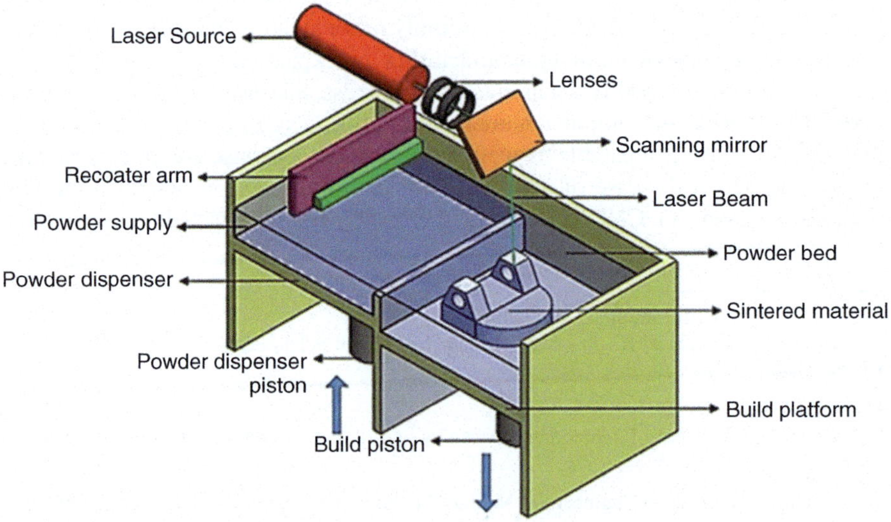

Fig. 8.1 A schematic illustration of DMLS process [10]

representation of the DMLS method. DMLS is a 3D printing technology that is based on laser, and it employs powdered metal, radiant heaters, and a computer-controlled laser to construct an object's one layer at a time. The machine fundamentally fabricates the object on a moving stage by spreading layers of the design material gradually. The machine applies a powder film with a precise thickness to each layer. Then, as shown in a computer-aided design (CAD) document, a powerful laser beam is coordinated on a powder bed and changed to interweave metal powders in its middle region, resulting in a thin metal layer. The stage descends the predetermined layer thickness, a new powder film is laid down, and the subsequent deposit is softened by exposure to the laser light, allowing it to adjust to the preceding layer [10]. As a result, the technique advances one layer after another until the component is built [11]. However, the DMLS technique takes around 6–12 h to create parts of even modest to moderate size [12].

DMLS is an additive manufacturing process that is a good alternative for producing parts with complex designs that machining or casting cannot produce in a timely manner. It provides a designer with the mechanical properties of aerospace materials as well as the freedom of design of 3D printing. Powdered metal is melted layer after layer using a powerful laser to create one or more designs based on a 3D CAD model. The printing of parts is done in an enclosed build chamber using argon and nitrogen gas infusion. In terms of density and strength, the end product of DMLS is comparable to a cast product right off the machine. However, hipping brings the part's strength characteristics very close to those of wrought parts [13].

DMLS is widely used in the aerospace industry. This technique integrated several assemblies into a single complicated item. For instance, the ArianeGroup effectively reduced a 248-component upper phase propulsion module into a single component using DMLS. The company achieved 50% cost-efficiency along with a significant reduction in production time by this way. Another good example with the DMLS is the Vectoflow, a company producing aerodynamic measurement devices. Vectoflow utilized the DMLS to produce a one-piece flow monitoring device that can sustain aircraft loads at subsonic and supersonic speeds. Moreover, General Electric effectively benefits from the DMLS method to manufacture Ti-6Al-4V structural airframe components including turbine blades. A liquid oxygen flange was additively manufactured for the upper part of United Launch Alliance launch vehicles, replacing traditional manufacturing methods [14–17]. In addition to the aerospace industry, additive manufacturing attracted much attention by the medical sector. Additive manufacturing technology including DMLS method allows for customized medical implants with porous structures and tailored mechanical properties to meet patient needs, such as custom hip implants [18]. Over the past few decades, there has been a significant increase in the demand for stamped automotive parts, which now exceeds 100 million per year. Stamped parts used in the automotive industry include chassis components such as A-pillars, B-pillars, roof rails, bumpers, tunnels, and rocker panels. DMLS can be used to manufacture sheet metal parts [19]. For the first few years of its commercial life, DMLS was primarily used to manufacture inserts for prototype injection molds ("rapid tooling"), with a small number of positive parts also built as prototype products ("rapid prototyping") [20].

DMLS process has the following advantages over conventional manufacturing processes: simple and sturdy process for complex parts with tooling elimination; reduction in the number of components for a given job with consequent weight reduction; good accuracy, precision, and surface finish with appropriate post-processing techniques; and reduction in cycle time. However, the challenges of DMLS process can be divided into two main groups: operational costs and process-related difficulties. First, limited availability of metal alloys in powder form for DMLS results in higher component costs. This process is currently limited to design prototyping. However, small complex parts such as fuel nozzles, swirlers, and cooled turbine blades are already mass-produced. Conventional machining can still be used for parts with simpler geometry and features. Regarding the process-related difficulties, limitations include size, shrinkage/warpage, and distortion. Minimizing parts involves dividing them into appropriate size sectors within the build envelope and designing suitable supports based on standard design rules [13].

8.2 Aeroengine Materials in DMLS

Aeroengines rely heavily on aerospace alloys, particularly superalloys, due to their exceptional properties. These alloys are chosen for their ability to withstand high temperatures; provide high strength and durability; resist corrosion, creep, and fatigue; and offer a lightweight design [21]. Critical components such as the turbine section and combustion chamber operate under extreme temperatures, demanding alloys engineered to maintain their integrity. Aeroengine alloys endure significant stress and mechanical strain, necessitating exceptional strength and durability to withstand operational pressures. Moreover, harsh conditions including combustion gases and environmental factors necessitate alloys with high corrosion resistance to ensure component durability [22]. Superalloys, engineered to resist creep deformation, are vital for components like turbine blades exposed to prolonged high temperatures during operation [23]. Fatigue, caused by cyclic loading, is another concern addressed in alloy selection to ensure long-term structural integrity under repetitive stress [24]. Despite alternative materials, alloys are preferred for their ability to produce lightweight components, crucial for enhancing fuel efficiency and overall performance in the aircraft industry [25].

To overcome all these harsh conditions during service, aeroengine alloys exhibit superior characteristics that make them difficult-to-process materials, especially in machining operations. For this reason, the DMLS technique can be used to build numerous aeroengine components as an alternative manufacturing process to the subtractive operations. The components include turbine blades, fuel nozzles, combustion chambers, exhaust components, heat exchangers, and various others. As commonly known, superalloys, mainly nickel-based ones, and titanium alloys are the main materials used to produce aeroengine components. In this section, DMLS process of nickel-based alloys and titanium alloys is discussed to understand the details about the processability of these aeroengine alloys via the DMLS technique.

8.2.1 Nickel-Based Alloys in DMLS

Because of their remarkable strength at high temperatures and resistance to corrosion, nickel-based alloys play a crucial role in the construction of all aerospace engines. Inconel 718 is one of the nickel-based alloys, being widely used in the production of aeroengine components. The utilization of Inconel 718 in additive manufacturing is recommended due to its exceptional durability, low thermal conductivity, and elevated hardness that hinder the processability of this alloy in conventional methods. The reason for this is that machined parts have a higher susceptibility to deterioration compared with additively made parts. The DMLS additively manufactured technology is very suitable for the production of this material [26].

Heat treatments are widely used to improve the mechanical properties of Inconel 718 components, such as yield and ultimate tensile strength, as well as to reduce porosity and residual stresses [27]. The commercial and DMLS alloys undergo 2 h of solutionizing at 1100 °C at a heat rate of 15 °C/min and then aging for 24 h at 845 °C [28]. Figure 8.2 shows the microstructure of the as-built DMLS specimen in the vertical plane. This image clearly depicts the laser scanning horizontally line by line and the melt pool with arc-shaped characteristics created vertically layer by layer. The melt ponds are shown in a greater detail with higher magnification (Fig. 8.2b). It is obvious that the metal has a totally dendritic structure established on the metal pool edge. When each track was deposited, the heat input and dissipation mode were consistent, encouraging the continuation of direct solidification. Figure 8.3 shows the recrystallization structure generated in the alloy following the heat treatment. It is also seen that the melt pool has completely vanished and a homogenized grain structure has developed in the alloy [29].

Kumar et al. [30] used various types of heat treatments: direct aged and solution aged for the additively manufactured Inconel 718 specimens. As-built specimen without any heat treatment was also analyzed as a reference in the comparison. It should be mentioned that no hot isostatic pressing (HIP) or stress reduction

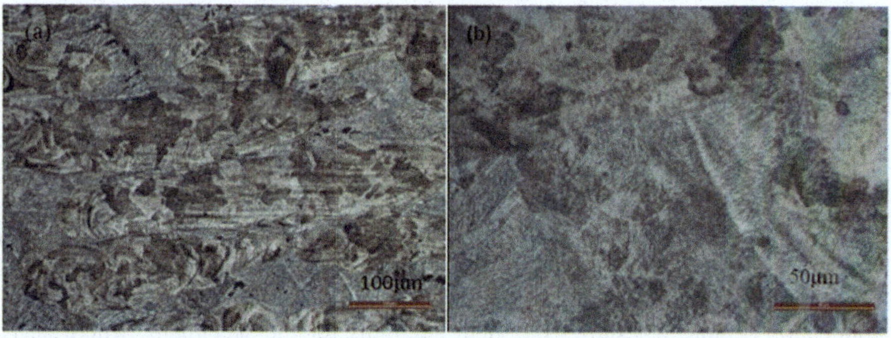

Fig. 8.2 Microscopy image of the DMLS sample before heat treatment: (**a**) higher magnification and (**b**) lower magnification [29]

Fig. 8.3 Microscopy image of the DMLS sample after heat treatment [29]

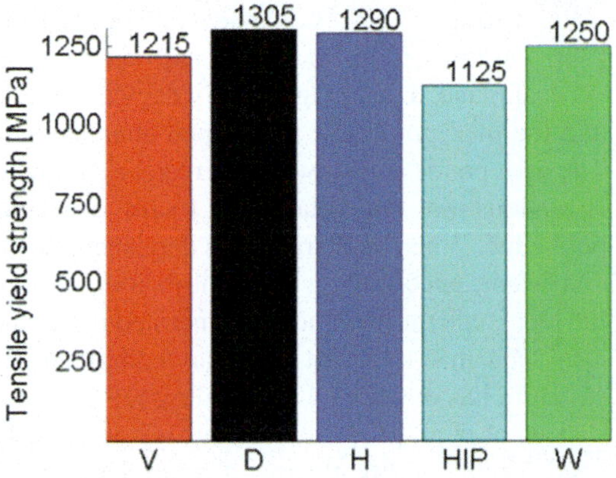

Fig. 8.4 Yield strengths for the specimens [3]

treatments were used in this investigation. Based on the results, heat treatments considerably enhanced the mechanical properties, such as increasing the ultimate tensile strength and yield strength when compared to as-built samples; however, heat treatment decreased the elongation at break. As-built alloy had an ultimate tensile strength of 996 MPa, whereas the ultimate tensile strengths were found as 1218 MPa and 1308 MPa for the solution-aged and direct-aged alloys, respectively. Smith et al. [3] investigated the microstructure and mechanical behavior of Inconel 718 produced via the DMLS method. The authors also applied HIP to the additively manufactured alloys to understand the microstructural evolution. The study reveals that the majority of porosity forms in the outer layer of the printed alloy. Nevertheless, porosity within the material itself is also evident. According to the mechanical tests, HIP lowers the yield strength of the alloy; however, ductility is enhanced by HIP application. Figure 8.4 shows the yield strengths for the specimens, where V, H, and D denote the vertically, horizontally, and diagonally built specimens, respectively, while HIP and W stand for HIPed and wrought specimens.

Gribbin et al. [31] investigated the low-cycle fatigue behavior of Inconel 718 specimens produced via the DMLS technique. Fatigue tests under a constant mean engineering strain of 0.5% and varying strain amplitudes from 0.6%–1.4% were conducted. It was observed that the fatigue performance of the additively manufactured specimen with heat treatment surpassed that of heat-treated wrought alloy at lower strain amplitudes. Conversely, the wrought alloy exhibited a longer lifespan at higher strain amplitudes. The findings indicate that the influence of porosity in the DMLS specimens is less pronounced at lower strain amplitudes compared with higher ones. Son et al. [32] investigated the creep behavior of additively manufactured Inconel 625. In the creep tests, the specimens were subjected to a range of stresses at 650–800 °C. Wrought alloys were also included in the tests as the reference group in addition to the HIPed specimens after the DMLS process. The ductility of HIPed alloys demonstrates a notable improvement over the non-HIPed alloys. However, the ductility of HIPed specimens remains significantly lower than that of the wrought alloy, especially with longer exposure times to test temperatures. The exact cause of the substantial decrease in ductility is not definitively pinpointed, but potential factors include sulfur segregation to grain boundaries and a relatively higher oxygen content, which could result in oxide embrittlement.

Raj et al. [33] studied the electrochemical behavior of DMLS-processed Inconel 718. To observe the heat treatment effect, a group of specimens were treated by solutionizing at 1100 °C for 2 h and aging at 845 °C for 24 h, followed by air-cooling. To investigate the corrosion behavior of the alloys, specimens were exposed to electrochemical polarization with 1.0 M H_2SO_4 as an electrolyte. From the microstructural and mechanical analyzes, as-built DMLS specimens showed anisotropic characteristics due to the layer-by-layer production. However, additively manufactured specimens changed their anisotropic behavior to isotropic properties after the heat treatment procedure. Moreover, additively manufactured alloys showed higher resistance to corrosion in comparison with the wrought alloys. The heat-treated wrought alloy exhibited intergranular corrosion in combination with pitting, having a high susceptibility to corrosion based on the corrosion tests. However, the heat-treated alloy produced via DMLS showed a higher resistance to corrosion, thereby protecting its structural integrity in aggressive corrosive medium. Figure 8.5 shows the corrosion morphologies on the specimens.

In addition to mechanical and chemical properties, researchers have also investigated the DMLS processing parameters for better surface characteristics in the parts. Hence, they made a great effort to eliminate the post-processing operations, such as surface machining or grinding. Uğur-Solakoğlu et al. [34] studied the DMLS process of Inconel 718 to enhance the surface finish of the additively manufactured products. In their study, laser power, scanning speed, and layer thickness were varied to optimize the processing parameters based on the surface roughness measurements. According to the measurements, surface roughness was suppressed by increasing laser power. On the other hand, lower scanning speeds and thinner layer depositions were suggested for reduced roughness on the product surface. Rifat et al. [35] investigated the centrifugal disc finishing operation for the additively manufactured Inconel 718 parts to increase the surface quality. They concluded the

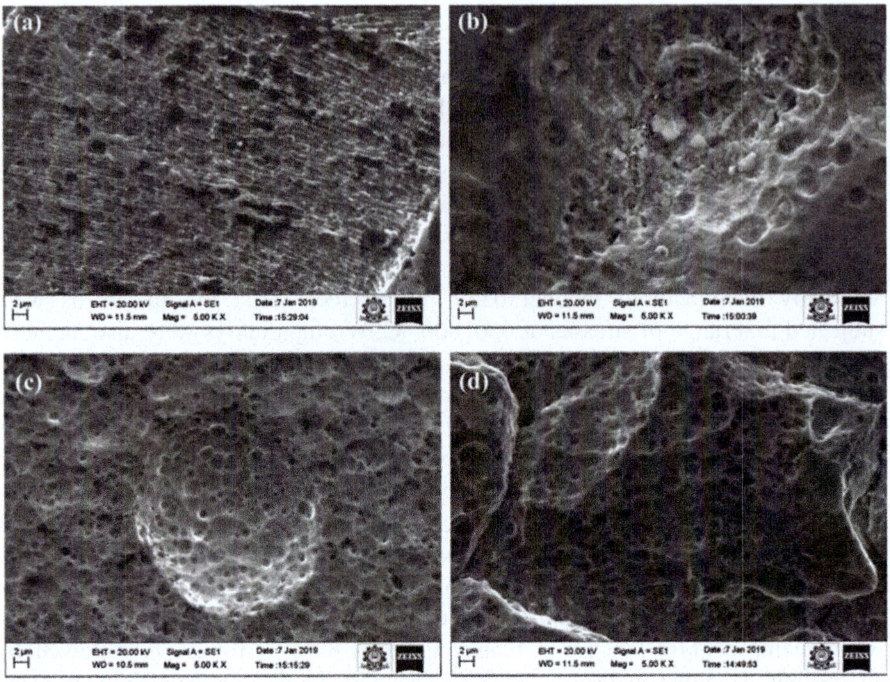

Fig. 8.5 Corrosion morphologies on the specimens: (**a**) DMLS, (**b**) heat-treated DMLS, (**c**) wrought, (**d**) heat-treated wrought [33]

study that additively manufactured alloys exhibit lower surface quality in comparison with the wrought alloys in their as-received states. Several unmelted powder particles are detected on the DMLS specimens due to the lack of fusion, thereby leading to a higher surface roughness in the final product. Furthermore, it is found that centrifugal disc finishing is an effective post-processing for lowering the roughness on the DMLS part surfaces. The improvement in the surface finish is directly related to the duration of the finishing operation as well as the process media.

8.2.2 Titanium Alloys in DMLS

Titanium alloys are extensively utilized in the aerospace sector to produce engine components, mainly the parts in low-pressure compressors. The most common titanium alloy in the aerospace industry is Ti-6Al-4V, which may reach a strength of over 1200 MPa after being quenched at 880–950 °C and tempered at 400–600 °C [36–38]. Because of its high strength-to-weight ratio, low density, strong corrosion resistance, and outstanding deformability, Ti-6Al-4V has become the most widely used titanium alloy in a variety of engineering applications [39, 40]. Therefore, a lot of airframes and engine parts are made from this alloy. This alloy has a wide range of real-world uses in aviation, where great reliability is necessary. Bolts, seat rails, and other similar components are utilized in airframes as general structural materials. In engines, the alloy is utilized for fan blades and fan casings and similarly in the intake section, where temperatures are relatively low, because of the comparatively low permissible temperature of roughly 300 °C [41].

Microstructures, internal defects, residual stresses, surface roughness, and other factors all affect materials' fatigue properties. These factors are influenced by the manufacturing procedures and post-manufacturing treatments used in the production. Furthermore, inclusions, vacancies within the material, surface roughness, and other defects can cause fatigue cracks. For this reason, actions must be taken to extend the fatigue life of additively manufactured metals, which have heterogeneous microstructures from inadequate melting and voids [42]. Post-heat treatments for additively manufactured Ti-6Al-4V do not significantly increase fatigue life because internal defects have a less severe effect on fracture initiation than on surface roughness [43]. In order to enhance the fatigue characteristics of additively manufactured titanium alloys, surface treatments are therefore a crucial technique [44]. The mechanical and microstructural qualities of Ti-6Al-4V components can be enhanced by heat treatment. Since titanium is an especially oxidation-sensitive material, the heat treatment is done in a vacuum or in an inert atmosphere [45].

Sallica-Leva et al. [46] studied the microstructure and mechanical properties of Ti-6Al-4V specimens produced via DMLS. They found that the alloy processed in the DMLS method had quite a good mechanics, comparable to the conventional Ti-6Al-4V alloys. Moreover, it showed higher mechanical properties than that produced in by the electron beam melting (EBM) technology. Brusa et al. [47] investigated the microstructure and mechanical properties properties of Ti-6Al-4V alloy produced in a DMLS process. Moreover, the DMLS method was compared with the EBM process in terms of the microstructural and mechanical behaviors of the alloys. Based on the tomographic analysis, the alloy produced in the EBM technique had several porosities in its microstructure, which drastically reduced the strength in the alloy. However, the DMLS method produced a dense microstructure in the alloy so that the mechanical properties were kept in a good range. Regarding the fatigue behavior, the EBM specimen showed a fracture at 729000 cycles, while the specimen produced via DMLS did not show a fracture up to 10^6 cycles. This was attributed to the denser microstructure of the DMLS alloy, which provided a good fatigue characteristic. On the other hand, DMLS produced a better surface finish in comparison with the EBM process (Fig. 8.6). Hence, crack initiation in the fatigue test was delayed at higher cycles in the DMLS specimen.

Greitemeier et al. [48] investigated the surface roughness of Ti-6Al-4V specimens produced in the DMLS method. In this study, roughness was tried to be suppressed to enhance the fatigue resistance of the alloy. The authors claimed that poor surface characteristics led to crack initiation at early fatigue cycles due to higher stress accumulations at the roots of the rough surface profiles. The DMLS process was applied at a laser power of 370 W while scanning the powder bed at a speed of 7000 mm/s. The thickness of the build-up layers was 60 μm in the process. Because

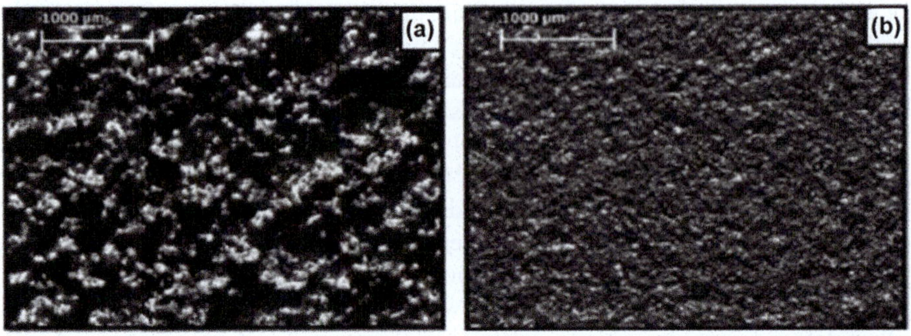

Fig. 8.6 Surface topography for the specimens produced via (**a**) EBM and (**b**) DMLS [47]

DMLS process leads to the development of residual stresses in the product, a heat treatment procedure was applied to the specimens to release the residual stresses. In this treatment, the specimens were rested at 710 °C for 2 h in a vacuum chamber; then, a furnace cooling was applied in an argon atmosphere. To avoid residual stress formations during cooling, gradual cooling at a slow rate was selected during the process. According to the roughness measurements, average surface roughness was 13 ± 0.7 µm on the specimens. The ultimate tensile and yield strengths were found as 1096 MPa and 1017 MPa, respectively, which are quite good results for the additively manufactured alloys. From the high-cycle fatigue results, fatigue properties were heavily dependent on surface roughness. This was concluded from the comparison of as-built and milled DMLS specimens that milled specimens showed higher fatigue resistance than the as-built ones. Moreover, the authors proposed a simple equivalent initial flaw size model for predicting the fatigue life of the DMLS alloys. It is also noteworthy that the proposed fatigue life estimation was a good approach to predict the effect of surface roughness.

Karolewska et al. [49] investigated the mechanical properties of Ti-6Al-4V alloys. The DMLS process was conducted with the following parameter settings: laser power of 200 W, minimum layer thickness of 30 µm, and scanning speed up to 7000 mm/s. The as-built specimens were then annealed to enhance the mechanical properties. The authors also included a wrought Ti-6Al-4V alloy as another specimen for comparison purpose. From the tensile testing results, the DMLS specimen showed a brittle characteristic along with higher strength compared with the wrought alloy. The ultimate tensile and yield strengths were 1127 MPa and 1052 MPa, respectively, for the DMLS alloy, whereas for the wrought alloy they were 1010 MPa and 947 MPa, respectively. However, elongations at break were 17% and 29% for the DMLS and wrought alloys, respectively. Similarly, hardness results were higher in the DMLS specimen compared with the wrought alloy. Figure 8.7 shows the hardness profiles for the specimens taken from the building direction (z-axis). In addition to these results, the authors indicated that the DMLS specimen showed anisotropic behavior due to the layer-by-layer building so that the properties could have some variations depending on the testing directions.

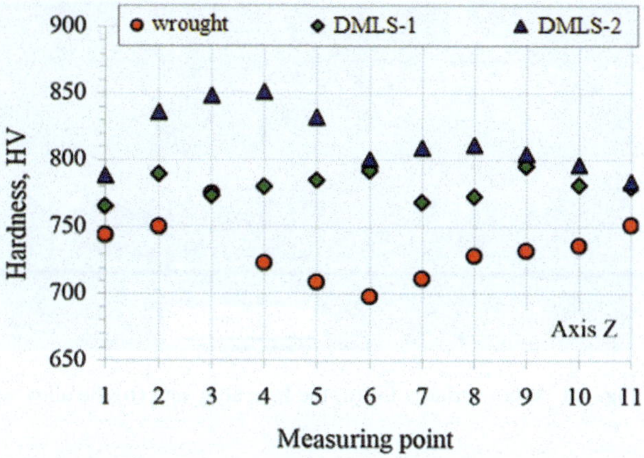

Fig. 8.7 Hardness profiles for the specimens [49]

Book et al. [50] investigated the strain localization in Ti-6Al-4V alloys produced via the DMLS method. From their microstructural investigation, β-phase boundaries play a critical role in the strain localization in the DMLS alloys. Hence, the authors stated that reducing the β-grain sizes during the DMLS process by applying higher cooling rates is beneficial to prevent β grain growth. The authors also emphasized that additively manufactured alloys have not proven their reliability due to the uncertainties in the alloy microstructures and mechanical properties. Therefore, critical components in the aerospace industry are still produced by using the conventional manufacturing methods despite the numerous advantages of additive manufacturing over the conventional ones.

Patil et al. [51] studied the wear resistance of Ti-6Al-4V alloy fabricated through the DMLS technique. Since high rotational speeds are used in the aeroengines, the components are subjected to severe frictional interactions at mating surfaces despite the heavy lubrication at these regions. The authors also benefited from TiB_2-reinforcing particles distributed in the titanium alloy powder for enhanced wear resistance. In the DMLS process, the parameters were selected as follows: laser power of 400 W, minimum laser spot size of 100 μm, and scanning speed of 200 mm/s. From the hardness measurements, reinforcing particles gradually increased the material hardness such that the bare alloy showed 383 HV, whereas the alloy with 2 wt% of TiB_2 particles had 477 HV. According to the tribological tests, particle reinforcement drastically enhanced the wear resistance of the alloy. For 600-s test duration, the bare alloy had about a wear thickness of 360 μm, while the alloy with 2 wt% of TiB_2 particles had only 250 μm. As shown in Fig. 8.8, the bare alloy shows microcracks and delamination on the worn surfaces, while the alloy having reinforcing particles has quite a smooth surface after the wear tests. This study showed that DMLS technology is also capable of producing particle reinforced alloys, such as metal matrix composites.

Wang et al. [52] investigated the repair of titanium-based aeroengine components by using additive manufacturing technology. They concentrated on the fatigue failures in the aeroengine parts made from titanium alloys and their characteristics

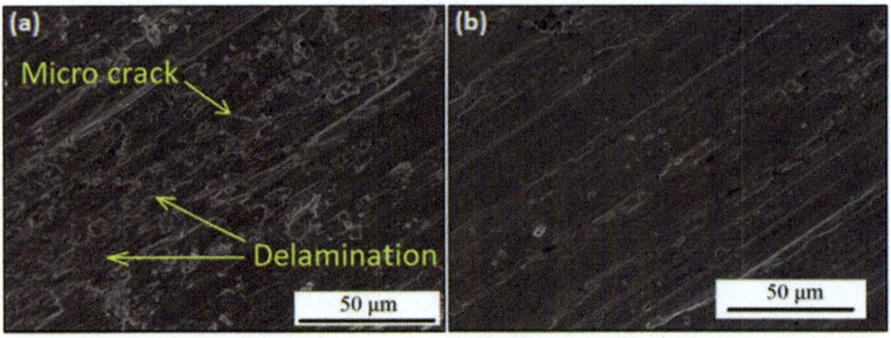

Fig. 8.8 Worn surfaces for (**a**) the bare alloy and (**b**) the alloy with 2 wt% of TiB_2 particles [51]

regarding repair processes. From this investigation, components in aeroengines are subjected to both low- and high-cycle fatigue conditions during their service life. Sometimes foreign object damage phenomenon leads to the crack initiation on the parts, thereby accelerating the fatigue progress in the alloys. Although some efforts have been made in the aerospace industry, the use of additive manufacturing is quite restricted for repair purposes. This is due to the fact that additive manufacturing for repair of titanium alloys provides lower plasticity and insufficient fatigue resistance despite good tensile and yield strengths achieved in the alloys. The main reason for reduced fatigue limit is the thermal residual stresses developed in the alloy during the additive manufacturing. For this reason, rotating components in aeroengines are generally not repaired by additive processes. However, various heat treatment designs after the additive manufacturing operations have promising results for enhanced fatigue properties in the repaired titanium alloys. Another challenge is porosity formation in the additive processes. However, HIP is effectively used to reduce the porosity in the microstructures. The authors also discussed the poor surface quality in additive repair operations. For this reason, grinding or any other surface processing is required after the additive repairing. Despite the promising results in additive repair processes for titanium alloys, the aerospace sector needs quite a long way to effectively use additive manufacturing as a repair technique.

8.3 Drawbacks and Future Outlook of DMLS in Aeroengine Applications

The DMLS is a very useful additive manufacturing process, and it is also an efficient substitute for machining or casting while producing parts with intricate designs. However, there are also some drawbacks in the DMLS method. One of them is its operational challenge. The limited supply of alloys of metals in powder form that are appropriate for the DMLS process and the need for more expensive components pose a challenge for this process. Thus, at this stage, this technique is only helpful for design prototyping. However, mass production has already begun for minor complex pieces such as swirlers, cooled turbine blades, and fuel nozzles. Parts with simpler geometry and properties are still capable of being machined successfully using conventional methods. Another drawback is the process-related challenges such as size constraints, warpage/shrinking, and deformation. By segmenting the part into sectors of appropriate dimensions within the built envelope and designing appropriate supports using established design standards these challenges can be minimized [53].

For aerospace applications, additive manufacturing technologies have the special benefit of producing difficult-to-machine materials including nickel alloys and titanium alloys. Nevertheless, the additive manufacturing industry lacks a common material characteristics database that would offer mechanical qualities, such as performance criteria and additional information on components fabricated by the DMLS method [54]. It is challenging for engineers to create without a thorough understanding of the characteristics of the materials that are utilized to design the components [55]. Comprehending and forecasting material attributes such as fatigue and surface roughness require the development of new physics-based additive manufacturing models [56], particularly for aerospace parts that operate in harsh conditions.

The certification process is a significant obstacle to the use of additive manufacturing in aerospace applications. The regulatory agencies must have assurance that additive manufacturing systems can be reliably built and tested to meet safety and reliability standards and that they are fundamentally understood thoroughly. The certification criteria for these systems differ based on their level of criticality, be it safety or mission-critical. Technical certification necessitates alignment with both developing standards for additive manufacturing processes and current requirements for traditional production [57]. According to the US Federal Aviation Regulations, several standards for commercial aircraft parts must be satisfied before a type certificate is issued [58]. This certification means that the design of the aircraft and all component parts (including propellers, engines, and control stations) are approved by the authority. As a result, it is necessary to evaluate the material effectiveness of even the most basic components and define performance variance. It might be necessary to conduct in-depth testing, collect data, and share to establish a high degree of confidence in the quality of components manufactured via additive manufacturing. The Federal Aviation Administration may grant a type certificate when the metallurgical process has been successfully qualified and the final product has been validated.

Although the DMLS still has a long way to predominate in the industry, it has strong potential for the future applications. Historically, the aerospace industry has used functional integration, topological optimization, and direct production to create complicated parts using novel materials. Aerospace applications mostly include designing, fabricating, and repairing aeroengine components using the DMLS technique [59]. This technology has the potential to create multifunctional or integrated aircraft components, simplifying the design and production process, eliminating fixtures and tools, and saving materials and fuel expenses. Integral aeroengine components eliminate joints, resulting in increased connection. The major goal is to replace connections such as bolts and screws, which undermine structural integrity and add significant cost [60]. Manufacturing on demand is the ability of the DMLS method to make items "just in time". Storage expenditures can be considerably lowered. On-demand fabrication ensures that parts remain undamaged by eliminating storage damage [61]. Printing components on demand in orbit would dramatically cut the cost and preparation time required to launch a rocket with substitution and maintenance equipment. Moreover, the aerospace industry has used the DMLS technique to repair expensive components including bearing housings, turbine blades, and blisks in aeroengines. This method saves money and minimizes the time required to fabricate and replace new components. Currently, the majority of time and money are spent on repair preparation. Automated repair techniques using the DMLS are predicted to successfully fix components in the future [59].

8.4 Conclusions

This study provided an overview of DMLS processing for aeroengine materials, focusing primarily on nickel-based and titanium alloys. It discussed material considerations, including microstructural and mechanical properties, while also providing details on the processing methods used in the DMLS. The DMLS method has the potential to transform the manufacturing system of aircraft engine components. The capabilities of DMLS to produce complex geometries with high accuracy are leading to a change in the traditional manufacturing methods of aerospace components. As the DMLS is an additive manufacturing process, material waste is minimal, further increasing its acceptability. However, like all other systems, the DMLS definitely comes with its drawbacks. The operational challenges of the DMLS and issues with surface finishing, microstructural porosities, and residual stresses are required to be solved in order to use it massively. In addition, components manufactured via the DMLS must adhere to certification rules and regulations for extensive utilization by aeroengine manufacturers.

References

1. Gibson I, Rosen DW, Stucker B. Additive manufacturing technologies: 3D printing, rapid prototyping and direct digital manufacturing. 2nd ed. New York and London: Springer; 2015.
2. Farber B, Small KA, Allen C, Causton RJ, Nichols A, Simbolick J, et al. Correlation of mechanical properties to microstructure in Inconel 718 fabricated by direct metal laser sintering. Mater Sci Eng A. 2018;712:539–47.
3. Smith DH, Bicknell J, Jorgensen L, Patterson BM, Cordes NL, Tsukrov I, et al. Microstructure and mechanical behavior of direct metal laser sintered Inconel alloy 718. Mater Charact. 2016;113:1–9.

4. Prakash KS, Nancharaih T, Rao VVS. Additive manufacturing techniques in manufacturing -an overview. Mater Today: Proc. 2018;5(2):3873–82.
5. Breaz RE, Bologa O, Racz SG. Selecting between CNC milling, robot milling and DMLS processes using a combined AHP and fuzzy approach. Procedia Comput Sci. 2017;122:796–803.
6. Sridhara V. Additive manufacturing of parts for indigenous aero engines. J Aerosp Sci Technol. 2017;69
7. Eosint M. Technical description EOSINT M 280 technical description technical description EOSINT M 280. EOS Gmbh; 2010.
8. Poyraz Ö, Yasa E, Akbulut G, Orhangül A, Pilatin S. Investigation of support structures for direct metal laser sintering (DMLS) of In625 Parts; 2015.
9. Scott-Emuakpor O, Schwartz J, George T, Holycross C, Cross C, Slater J. Bending fatigue life characterisation of direct metal laser sintering nickel alloy 718. Fatigue Fract Eng Mat Struct. 2015 Sep;38(9):1105–17.
10. Srinivasan D, Meignanamoorthy M, Ravichandran M, Mohanavel V, Alagarsamy SV, Chanakyan C, et al. 3D printing manufacturing techniques, materials, and applications: an overview. In: Rambo R, editor. Advances in materials science and engineering; 2021. p. 1–10.
11. Mangano F, Chambrone L, Van Noort R, Miller C, Hatton P, Mangano C. Direct metal laser sintering titanium dental implants: a review of the current literature. Int J Biomater. 2014;2014: 1–11.
12. Verma A, Tyagi S, Yang K. Modeling and optimization of direct metal laser sintering process. Int J Adv Manuf Technol. 2015;77(5–8):847–60.
13. Sridhara V. Additive manufacturing of parts for indigenous aero engines. J Aerosp Sci Technol. 2023;69(1A):191–4.
14. Schiller GJ. Additive manufacturing for aerospace. In: 2015 IEEE aerospace conference; 2015. p. 1–8.
15. ArianeGroup. Future ariane propulsion module: simplified by additive manufacturing. EOS Gmbh; 2017.
16. Vectoflow. Durable up to the sound barrier and beyond. EOS Gmbh; 2017.
17. Arcam A. EBM® electron beam melting-in the forefront of additive manufacturing. Arcam EBM, http://www arcam com/technology/electron-beam-melting/; 2019. p. 5.
18. Vranić A, Bogojevic N, Ćirić Kostić S, Croccolo D, Olmi G. Advantages and drawbacks of additive manufacturing; 2017.
19. Leal R, Barreiros FM, Alves L, Romeiro F, Vasco JC, Santos M, et al. Additive manufacturing tooling for the automotive industry. Int J Adv Manuf Technol. 2017;92(5–8):1671–6.
20. Shellabear M, Nyrhilä O. DMLS-Development history and state of the art. In: Laser assisted netshape engineering 4, proceedings of the 4th LANE; 2004. p. 21–4.
21. Geddes B, Leon H, Huang X. Superalloys: alloying and performance. ASM International; 2010.
22. Gialanella S, Malandruccolo A. Aerospace alloys. Springer; 2020.
23. Reed RC. The superalloys: fundamentals and applications. Cambridge University Press; 2008.
24. Stephens RI, Fatemi A, Stephens RR, Fuchs HO. Metal fatigue in engineering. Wiley; 2000.
25. Zhu L, Li N, Childs P. Light-weighting in aerospace component and system design. Propuls Power Res. 2018;7(2):103–19.
26. Keshavarzkermani A, Sadowski M, Ladani L. Direct metal laser melting of Inconel 718: process impact on grain formation and orientation. J Alloys Compd. 2018;736:297–305.
27. Teixeira Ó, Silva FJ, Atzeni E. Residual stresses and heat treatments of Inconel 718 parts manufactured via metal laser beam powder bed fusion: an overview. Int J Adv Manuf Technol. 2021;113(11–12):3139–62.
28. El-Bagoury N, Amin MA, Mohsen Q. Effect of various heat treatment conditions on microstructure, mechanical properties and corrosion behavior of Ni base superalloys. Int J Electrochem Sci. 2011;6(12):6718–32.
29. Raj BA, Jappes JW, Khan MA, Dillibabu V, Brintha N. Direct metal laser sintered (DMLS) process to develop Inconel 718 alloy for turbine engine components. Optik. 2020;202:163735.

30. Kumar MA, Amit K. Study of microstructure and mechanical properties of as-built and heat-treated additive manufactured inconel 718 alloy. ASEAN J Sci Technol Dev. 2021;38(3):6.
31. Gribbin S, Bicknell J, Jorgensen L, Tsukrov I, Knezevic M. Low cycle fatigue behavior of direct metal laser sintered Inconel alloy 718. Int J Fatigue. 2016;93:156–67.
32. Son K, Kassner ME, Lee KA. The creep behavior of additively manufactured Inconel 625. Adv Eng Mater. 2020;22(1):1900543.
33. Raj BA, Jappes JTW, Khan MA, Dillibabu V, Brintha NC. Studies on heat treatment and electrochemical behaviour of 3D printed DMLS processed nickel-based superalloy. Appl Phys A Mater Sci Process. 2019;125(10):722.
34. Uğur Solakoğlu E, Gürgen S, Kuşhan MC. Surface topography of nickel-based superalloy manufactured with direct metal laser sintering (DMLS) method. Surf Topogr: Metrol Prop. 2019;7(1):015012.
35. Rifat M, DeMeter E, Basu S. Evolution of surface roughness of wrought and additively manufactured Inconel 718 during centrifugal disc finishing. JOM. 2024;76:2835.
36. Filip R, Kubiak K, Ziaja W, Sieniawski J. The effect of microstructure on the mechanical properties of two-phase titanium alloys. J Mater Process Technol. 2003;133(1–2):84–9.
37. Filip R, Pleshakov E. The effects of laser nitriding on improving the properties of machine elements made of titanium alloys. Adv Manuf Sci Technol. 2004;28(1):5–22.
38. Sieniawski J. Nickel and titanium alloys in aircraft turbine engines. Adv Manuf Sci Technol. 2003;27(3):23–33.
39. Sibum H. Titanium and titanium alloys—from raw material to semi-finished products. Adv Eng Mater. 2003;5(6):393–8.
40. Ao N, Liu D, Liu C, Zhang X, Liu D. Face-centered titanium induced by ultrasonic surface rolling process in Ti-6Al-4V alloy and its tensile behavior. Mater Charact. 2018;145:527–33.
41. Inagaki I, Takechi T, Shirai Y, Ariyasu N. Application and features of titanium for the aerospace industry. Nippon Steel Sumitomo Metal Techn Rep. 2014;106(106):22–7.
42. Pegues J, Shao S, Shamsaei N, Sanaei N, Fatemi A, Warner D, et al. Fatigue of additive manufactured Ti-6Al-4V, part I: the effects of powder feedstock, manufacturing, and post-process conditions on the resulting microstructure and defects. Int J Fatigue. 2020;132:105358.
43. Kahlin M, Ansell H, Moverare J. Fatigue behaviour of notched additive manufactured Ti6Al4V with as-built surfaces. Int J Fatigue. 2017;101:51–60.
44. Kuji C, Soyama H. Mechanical surface treatment of titanium alloy Ti6Al4V manufactured by direct metal laser sintering using laser cavitation. Metals. 2023;13(1):181.
45. Wanying L, Yuanhua L, Yuhai C, Taihe S, Singh A. Effect of different heat treatments on microstructure and mechanical properties of Ti6Al4V titanium alloy. Rare Metal Mater Eng. 2017;46(3):634–9.
46. Sallica-Leva E, Jardini AL, Fogagnolo JB. Microstructure and mechanical behavior of porous Ti–6Al–4V parts obtained by selective laser melting. J Mech Behav Biomed Mater. 2013;26: 98–108.
47. Brusa E, Sesana R, Ossola E. Numerical modeling and testing of mechanical behavior of AM titanium alloy bracket for aerospace applications. Procedia Struct Integr. 2017;5:753–60.
48. Greitemeier D, Dalle Donne C, Syassen F, Eufinger J, Melz T. Effect of surface roughness on fatigue performance of additive manufactured Ti–6Al–4V. Mater Sci Technol. 2016;32(7): 629–34.
49. Karolewska K, Ligaj B, Wirwicki M, Szala G. Strength analysis of Ti6Al4V titanium alloy produced by the use of additive manufacturing method under static load conditions. J Mater Res Technol. 2020;9(2):1365–79.
50. Book TA, Sangid MD. Strain localization in Ti-6Al-4V Widmanstätten microstructures produced by additive manufacturing. Mater Charact. 2016;122:104–12.
51. Patil AS, Hiwarkar VD, Verma PK, Khatirkar RK. Effect of TiB2 addition on the microstructure and wear resistance of Ti-6Al-4V alloy fabricated through direct metal laser sintering (DMLS). J Alloys Compd. 2019;777:165–73.

52. Wang L, Li Y, Zhou L, Lou Y, Liu S, Zheng D, et al. Progress in additive manufacturing, additive repair and fatigue evaluation of aviation titanium alloy blades. Mater Res Lett. 2023;11 (12):973–1012.
53. Sridhara V. Additive manufacturing of parts for indigenous aero engines. J Aerosp Sci Technol. 2017:191–4.
54. Campbell T, Williams C, Ivanova O, Garrett B. Could 3D printing change the world. In: Technologies, potential, and implications of additive manufacturing, vol. 3. Washington: Atlantic Council; 2011. p. 1–16.
55. Gupta N, Weber C, Newsome S. Additive manufacturing: status and opportunities. Washington: Science and Technology Policy Institute; 2012.
56. Frazier WE. Direct digital manufacturing of metallic components: vision and roadmap. In: 2010 international solid freeform fabrication symposium. University of Texas at Austin; 2010.
57. Blakey-Milner B, Gradl P, Snedden G, Brooks M, Pitot J, Lopez E, et al. Metal additive manufacturing in aerospace: a review. Mater Des. 2021;209:110008.
58. Federal Aviation Administration. Aircraft Certification [Internet]; 2016. Available from: https://www.faa.gov/aircraft/air_cert
59. Liu R, Wang Z, Sparks T, Liou F, Newkirk J. Aerospace applications of laser additive manufacturing. In: Laser additive manufacturing. Elsevier; 2017. p. 351–71.
60. Gausemeier J, Echterhoff N, Kokoschka M, Wall M. Thinking ahead the future of additive manufacturing. In: Future applications; 2011.
61. Bourell DL, Leu MC, Rosen DW. Roadmap for additive manufacturing: identifying the future of freeform processing. University of Texas at Austin Labratory for Freeform Fabrication Advanced Manufacturing Center; 2009.

Index

S. Gürgen (ed.), *Advances in Manufacturing for Aerospace Alloys*, Sustainable Aviation, https://doi.org/10.1007/978-3-031-64455-9

The manufacturer's authorised representative in the EU is Springer Nature Customer Service Centre GmbH, Europaplatz 3, 69115 Heidelberg, Germany. If you have any concerns regarding our products, please contact ProductSafety@springernature.com

Printed and bound by CPI Group (UK) Ltd, Croydon, CR0 4YY
01/10/2026
02225919-0015